W9-CFV-742

# teach yourself ®

## mathematics
### trevor johnson
### and hugh neill

teach yourself
70
1938
2008
celebrate
with us

Launched in 1938, the **teach yourself** series grew rapidly in response to the world's wartime needs. Loved and trusted by over 50 million readers, the series has continued to respond to society's changing interests and passions and now, 70 years on, includes over 500 titles, from Arabic and Beekeeping to Yoga and Zulu. What would you like to learn?

be where you want to be with **teach yourself**

For UK order enquiries: please contact Bookpoint Ltd, 130 Milton Park, Abingdon, Oxon OX14 4SB. Telephone: +44 (0) 1235 827720. Fax: +44 (0) 1235 400454. Lines are open 09.00–17.00, Monday to Saturday, with a 24-hour message answering service. Details about our titles and how to order are available at www.teachyourself.co.uk

For USA order enquiries: please contact McGraw-Hill Customer Services, PO Box 545, Blacklick, OH 43004-0545, USA. Telephone: 1-800-722-4726. Fax: 1-614-755-5645.

For Canada order enquiries: please contact McGraw-Hill Ryerson Ltd, 300 Water St, Whitby, Ontario L1N 9B6, Canada. Telephone: 905 430 5000. Fax: 905 430 5020.

Long renowned as the authoritative source for self-guided learning – with more than 50 million copies sold worldwide – the **teach yourself** series includes over 500 titles in the fields of languages, crafts, hobbies, business, computing and education.

*British Library Cataloguing in Publication Data*: a catalogue record for this title is available from the British Library.

*Library of Congress Catalog Card Number:* on file.

First published in UK 2001 by Hodder Education, part of Hachette Livre UK, 338 Euston Road, London, NW1 3BH.

First published in US 2001 by The McGraw-Hill Companies, Inc.

This edition published 2008.

The **teach yourself** name is a registered trade mark of Hodder Headline.

Typeset by Macmillan India Limited
Printed in Great Britain for Hodder Education, an Hachette Livre UK Company, 338 Euston Road, London NW1 3BH, by CPI Cox & Wyman, Reading, Berkshire, RG1 8EX.

The publisher has used its best endeavours to ensure that the URLs for external websites referred to in this book are correct and active at the time of going to press. However, the publisher and the author have no responsibility for the websites and can make no guarantee that a site will remain live or that the content will remain relevant, decent or appropriate.

Hachette Livre UK's policy is to use papers that are natural, renewable and recyclable products and made from wood grown in sustainable forests. The logging and manufacturing processes are expected to conform to the environmental regulations of the country of origin.

Impression number   10 9 8 7 6 5 4 3 2 1

Year                      2012 2011 2010 2009 2008

# contents

# preface

*Teach Yourself Mathematics* aims to give the reader a broad mathematical experience and a firm foundation for further study. It reflects recent developments in the subject, including, for example, chapters on both Probability and Statistics, which now take a more central position in mathematics. In this new edition, the authors have taken the opportunity to bring some questions up to date, especially those involving pre-euro European currencies.

This book is primarily aimed at the student who does not have teacher support. Consequently, the explanations are detailed and there are numerous worked examples. The book will also be a useful source of reference for homework or revision for students who are studying a mathematics course.

Access to a calculator has been assumed throughout the book. Generally, a basic calculator is adequate but, for Chapters 20 and 21, a scientific calculator, that is, a calculator which includes the trigonometric functions of sine, cosine and tangent, is essential. Very little else, other than some knowledge of basic arithmetic, has been assumed.

To derive maximum benefit from *Teach Yourself Mathematics*, work through and don't just read through the book. This applies not only to the exercises but also to the examples. Doing this will help your mathematical confidence grow.

The authors would like to thank the staff of Hodder Education for the valuable advice they have given.

Trevor Johnson, Hugh Neill

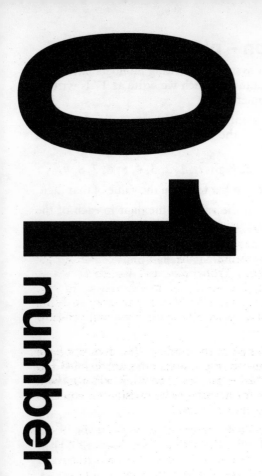

# 01 number

In this chapter you will learn:

- about place value
- about the four operations of arithmetic
- about the order in which arithmetic operations should be carried out
- about some special numbers.

# 1.1 Introduction – place value

Over the centuries, many systems of writing numbers have been used. For example, the number which we write as 17 is written as XVII using Roman numerals.

Our system of writing numbers is called the **decimal system**, because it is based on ten, the number of fingers and thumbs we have.

The decimal system uses the **digits** 0, 1, 2, 3, 4, 5, 6, 7, 8, 9.

The place of the digit in a number tells you the value of that digit.

When you write a number, the value of the digit in each of the first four columns, starting from the right, is

| thousands | hundreds | tens | units |
|:---:|:---:|:---:|:---:|
| 3 | 6 | 5 | 2 |

Thus, in the number 3652, the value of the 2 is 2 units and the value of the 6 is 6 hundreds.

The decimal system uses zero to show that a column is empty. So 308 means 3 hundreds and 8 units, that is, there are no tens. The number five thousand and twenty three would be written 5023, using the zero to show that there are no hundreds.

Extra columns may be added to deal with larger numbers, the value of each column being ten times greater than that of the column on its immediate right. It is usual, however, to split large numbers up into groups of three digits.

You read 372 891 as three hundred and seventy two thousand, eight hundred and ninety one.

In words, 428 763 236 is four hundred and twenty eight million, seven hundred and sixty three thousand, two hundred and thirty six.

## EXERCISE 1.1

1 What is the value of the 5 in each of the following numbers?
   (a) 357   (b) 598   (c) 5842   (d) 6785
2 Write these numbers in figures.
   (a) seventeen          (b) seventy
   (c) ninety seven       (d) five hundred and forty six
   (e) six hundred and three  (f) eight hundred and ten
   (g) four hundred and fifty  (h) ten thousand

    (i) eight thousand, nine hundred and thirty four
    (j) six thousand, four hundred and eighty
    (k) three thousand and six

**3** Write these numbers completely in words.
    (a) 52    (b) 871    (c) 5624    (d) 980
    (e) 7001    (f) 35 013    (g) 241 001    (h) 1 001 312

**4** Write down the largest number and the smallest number you can make using
    (a) 5, 9 and 7,    (b) 7, 2, 1 and 9.

**5** Write these numbers completely in words.
    (a) 342 785    (b) 3 783 194
    (c) 17 021 209    (d) 305 213 097

**6** Write these numbers in figures.
    (a) five hundred and sixteen thousand, two hundred and nineteen
    (b) two hundred and six thousand and twenty four
    (c) twenty one million, four hundred and thirty seven thousand, eight hundred and sixty nine
    (d) seven million, six hundred and four thousand and thirteen

# 1.2 Arithmetic – the four operations

In this book, the authors assume that you have access to a calculator, and that you are able to carry out simple examples of the four operations of arithmetic, addition, subtraction, multiplication and division, either in your head or on paper. If you have difficulty with this, you may wish to consult *Teach Yourself Basic Mathematics*.

Each of the operations has a special symbol, and a name for the result.

Adding 4 and 3 is written $4 + 3$, and the result, 7, is called the **sum** of 4 and 3.

Subtracting a number, 4, from a larger number, 7, is written $7 - 4$, and the result, 3, is their **difference**.

Addition and subtraction are 'reverse' processes:
    $4 + 3 = 7$ and $7 - 4 = 3$.

Multiplying 8 by 4 is written $8 \times 4$, and the result, 32, is called the **product** of 8 by 4.

Dividing 8 by 4 is written $8 \div 4$, or $\frac{8}{4}$, and the result, 2, is called their **quotient**.

Multiplication and division are 'reverse' processes:
$4 \times 2 = 8$ and $8 \div 4 = 2$.

Sometimes divisions are not exact. For example, if you divide 7 by 4, the quotient is 1 and the **remainder** is 3, because 4 goes into 7 once and there is three left over.

Exercise 1.2 gives you practice with the four operations. You should be able to do all of this exercise without a calculator.

### EXERCISE 1.2

1 Work out the results of the following additions.
   (a) $23 + 6$   (b) $33 + 15$   (c) $45 + 9$   (d) $27 + 34$

2 Find the sum of
   (a) 65, 24 and 5,      (b) 27, 36 and 51.

3 Work out the following differences.
   (a) $73 - 9$      (b) $49 - 35$      (c) $592 - 76$
   (d) $128 - 43$   (e) $124 - 58$   (f) $171 - 93$

4 Find the difference between
   (a) 152 and 134,      (b) 317 and 452.

5 Work out
   (a) $5 \times 9$   (b) $43 \times 7$   (c) $429 \times 6$   (d) $508 \times 7$
   (e) $27 \times 11$   (f) $12 \times 28$   (g) $13 \times 14$   (h) $21 \times 21$

6 Find the product of 108 and 23.

7 Work out
   (a) $375 \div 5$   (b) $846 \div 3$   (c) $1701 \div 7$   (d) $1752 \div 8$

8 Find the quotients and the remainders for the following divisions.
   (a) $8 \div 3$   (b) $23 \div 5$   (c) $101 \div 8$

## 1.3 Order of operations

Sometimes you will see a calculation in the form $8 - (2 \times 3)$, where brackets, that is the symbols (and), are put around part of the calculation. Brackets are there to tell you to carry out the calculation inside them first. So $8 - (2 \times 3) = 8 - 6 = 2$.

If you had to find the answer to $8 + 5 - 3$ , you might wonder whether to do the addition first or the subtraction first.

If you do the addition first, $8 + 5 - 3 = 13 - 3 = 10$.

If you do the subtraction first, $8 + 5 - 3 = 8 + 2 = 10$.

In this case, it makes no difference which operation you do first.

If, however, you had to find the answer to $3 + 4 \times 2$, your answer will depend on whether you do the addition first or the multiplication first.

If you do the addition first, $3 + 4 \times 2 = 7 \times 2 = 14$.

If you do the multiplication first, $3 + 4 \times 2 = 3 + 8 = 11$.

To avoid confusion, mathematicians have decided that, if there are no brackets, multiplication and division are carried out before addition and subtraction. Such a rule is called a **convention**: clearly the opposite decision could be made, but it has been agreed that the calculation $3 + 4 \times 2$ should be carried out as $3 + 4 \times 2 = 3 + 8 = 11$.

There are a number of these conventions, often remembered by the word 'BoDMAS', which stands for
    Brackets,
    Divide,
    Multiply,
    Add,
    Subtract.

This is the order in which arithmetic operations must be carried out. Most calculators have the 'BoDMAS' convention programmed into them, and so automatically carry out operations in the correct order.

---

**Example 1.3.1**

Carry out the following calculations
(a) $5 \times (6 - 2)$    (b) $(10 + 2) \div 4$    (c) $20 - 3 \times 5$
(d) $24 \div 6 + 2$    (e) $4 \times (8 - 3) + 2$

(a) $5 \times (6 - 2) = 5 \times 4$        brackets first
$\qquad\qquad\quad = 20$

(b) $(10 + 2) \div 4 = 12 \div 4$      brackets first
$\qquad\qquad\qquad = 3$

(c) $20 - 3 \times 5 = 20 - 15$      multiply before subtracting
$\qquad\qquad\qquad = 5$

(d) $24 \div 6 + 2 = 4 + 2 = 6$     division before addition

(e) $4 \times (8 - 3) + 2 = 4 \times 5 + 2$   brackets first
$\qquad\qquad\qquad\quad = 20 + 2$    multiply before adding
$\qquad\qquad\qquad\quad = 22$

Try the calculations in Example 1.3.1 on your calculator to see if it uses the 'BoDMAS' convention and carries them out correctly.

### EXERCISE 1.3

In each part of Question 1, use the BoDMAS convention to carry out the calculation.

1 (a) $(3+4) \times 5$      (b) $3+4 \times 5$      (c) $(7-3) \times 2$

  (d) $7-3 \times 2$        (e) $(12 \div 6) \div 2$      (f) $12 \div (6 \div 2)$

  (g) $20 \div 4 + 1$       (h) $3 \times 5 + 4 \times 2$    (i) $4 \times (7-2) + 6$

  (j) $6 \times 4 - 2 \times 8$     (k) $24 \div (4+2) - 1$ (l) $(7+3) \times (9-2)$

  (m) $10 + (8+4) \div 3$ (n) $9 + 5 \times (8-2)$  (o) $15 \div 3 - 16 \div 8$

  (p) $5 + 6 \times 3 - 1$    (q) $(5+6) \times 3 - 1$  (r) $5 + 6 \times (3-1)$

## 1.4 Problems which use arithmetic

In Exercise 1.4, you have to decide which arithmetic operation to use. If you are not sure, it can sometimes be helpful to think of a similar question with smaller numbers.

---

**Example 1.4.1**

An egg box holds six eggs. How many eggs boxes are needed to pack 500 eggs, and how many eggs are there in the last box?

You need to find how many sixes there are in 500. If you use a calculator you get $500 \div 6 = 83.333\,333\,3$.

This means that 83 boxes will be full. These 83 full boxes hold $83 \times 6$ eggs, which is 498 eggs. So there are 2 eggs in the last box.

---

### EXERCISE 1.4

1 In a school, 457 of the pupils are boys and 536 are girls. Work out the number of pupils in the school.

2 How many hours are there in 7 days?

3 The sum of two numbers is 3472. One of the numbers is 1968. Find the other number.

4 The product of two numbers is 161. One of the numbers is 7. Find the other number.

5 A car travels 34 miles on one gallon of petrol. How far will it travel on 18 gallons?

6 How many pieces of metal 6 cm long can be cut from a bar 117 cm long? What length of metal is left over?

7 A box holds 24 jars of coffee. How many boxes will be needed to hold 768 jars?

8 At the start of a journey, a car's milometer read 34 652. After the journey, it read 34 841. How long was the journey?

9 There are 365 days in a year. How many days are there in 28 years?

10 A coach can carry 48 passengers. How many coaches are needed to carry 1100 passengers? How many seats will be unoccupied?

# 1.5 Special numbers

The numbers 0, 1, 2, 3, 4, ... are called **whole numbers**.

The numbers 1, 2, 3, 4, ... are called **natural numbers**. They are also called **counting numbers** or **positive whole numbers**.

The numbers ..., −3, −2, −1, 0, 1, 2, 3, ... are called **integers**. These are the positive and negative whole numbers, together with 0.

The natural numbers can be split into even and odd numbers.

An **even number** is one into which 2 divides exactly. The first six even numbers are 2, 4, 6, 8, 10 and 12. You can tell whether a number is even by looking at its last digit. If the last digit is 2, 4, 6, 8 or 0, then the number is even. The last digit in the number 784 is 4, so 784 is even.

An **odd number** is one into which 2 does not divide exactly, that is, it is a natural number which is not even. The first six odd numbers are 1, 3, 5, 7, 9 and 11. You can tell whether a number is odd by looking at its last digit. If the last digit is 1, 3, 5, 7 or 9, then the number is odd. In other words, if the last digit is odd, then the number is odd. The last digit in the number 837 is odd, so 837 is odd.

A **square number** is a natural number multiplied by itself. It is sometimes called a **perfect square** or simply a **square**, and can be shown as a square of dots. The first four square numbers are $1 \times 1 = 1$, $2 \times 2 = 4$, $3 \times 3 = 9$ and $4 \times 4 = 16$ (see Figure 1.1).

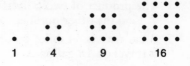

**Figure 1.1**

To get a square number you multiply a whole number by itself. So the fifth square number is $5 \times 5 = 25$.

A **triangle number** can be illustrated as a triangle of dots (see Figure 1.2). The first four triangle numbers are 1, 3, 6 and 10.

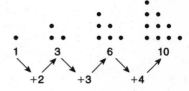

**Figure 1.2**

Each triangle number is obtained from the previous one by adding one more row of dots down the diagonal. The fifth triangle number is found by adding 5 to the fourth triangle number, so the fifth triangle number is 15.

A **rectangle number** can be shown as a rectangle of dots. A single dot or a line of dots is not regarded as a rectangle. Figure 1.3 shows the rectangle number 12 illustrated in two ways.

**Figure 1.3**

A **cube number** can be shown as a cube of dots (see Figure 1.4). The first three cube numbers are 1, 8 and 27 which can be written $1 = 1 \times 1 \times 1$, $8 = 2 \times 2 \times 2$ and $27 = 3 \times 3 \times 3$. To get a cube number you multiply a whole number by itself, and then by itself again. So the fourth cube number is $4 \times 4 \times 4 = 64$.

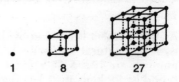

**Figure 1.4**

## EXERCISE 1.5

1 From the list 34, 67, 112, 568, 741 and 4366, write down
   (a) all the even numbers,
   (b) all the odd numbers.

2 Write down the first ten square numbers.

3 Write down the first ten triangle numbers.

4 Write down the first ten rectangle numbers.

5 Write down the first ten cube numbers.

6 From the list of numbers 20, 25, 28, 33, 36 and 37, write down
   (a) all the square numbers,
   (b) all the triangle numbers,
   (c) all the rectangle numbers.

7 The difference between consecutive square numbers is always the same type of number. What type of number is it?

8 The sum of the first two triangle numbers is $4 = (1 + 3)$.
   (a) Find the sums of three more pairs of consecutive triangle numbers.
   (b) What do you notice about your answers?

9 Find the sums of the first two, three, four and five cube numbers. What do you notice about your answers?

10 (a) Write down a rectangle number, other than 12, for which there are two different patterns of dots. Draw the patterns.
   (b) Write down a rectangle number for which there are three different patterns of dots.
   (c) Write down a rectangle number for which there are four different patterns of dots.

# 1.6 Multiples, factors and primes

Figure 1.5 shows the first five numbers in the 8 times table. The numbers on the right, 8, 16, 24, 32 and 40, are called **multiples** of 8. Notice that 8 itself, which is $1 \times 8$ or $8 \times 1$ is a multiple of 8.

$$1 \times 8 = 8$$
$$2 \times 8 = 16$$
$$3 \times 8 = 24$$
$$4 \times 8 = 32$$
$$5 \times 8 = 40$$

**Figure 1.5**

The **factors** of a number are the numbers which divide exactly into it. So, the factors of 18 are 1, 2, 3, 6, 9 and 18. Notice that 1 is a factor of every number and that every number is a factor of itself.

### Example 1.6.1

Write down all the factors of (a) 12, (b) 11.
(a) The factors of 12 are 1, 2, 3, 4, 6 and 12.
(b) The factors of 11 are 1 and 11.

The number 11 has only two factors and is an example of a prime number. A **prime number**, sometimes simply called a **prime**, is a number greater than 1 which has exactly two factors, itself and 1.

Thus, the first eight prime numbers are 2, 3, 5, 7, 11, 13, 17 and 19.

## EXERCISE 1.6

1 Write down the first five multiples of 7.

2 (a) Write down the first ten multiples of 5.
  (b) How can you tell that 675 is also a multiple of 5?

3 Write down all the factors of 20.

4 (a) Find the sum of the factors of 28, apart from 28 itself.
  (b) Comment on your answer.

5 Write down all the factors of 23.

6 (a) The number 4 has three factors (1, 2 and 4). Find five more numbers, each of which has an odd number of factors.
  (b) What do you notice about your answers to part (a)?

7 Write down the four prime numbers between 20 and 40.

8 2 is the only even prime number. Explain why there cannot be any others.

9 Write down the largest number which is a factor of 20 and is also a factor of 30.

10 Write down the smallest number which is a multiple of 6 and is also a multiple of 8.

11 Write down the factors of 30 which are also prime numbers. (These are called the prime factors of 30.)

12 Write down the prime factors of 35.

13 The prime factors of 6 are 2 and 3. Find another number which has 2 and 3 as its only prime factors.

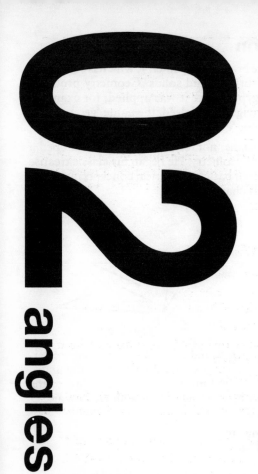

# 02

# angles

**In this chapter you will learn:**

- about different types of angles
- how to measure and draw angles
- angle facts and how to use them
- about parallel lines
- about bearings.

# 2.1 Introduction

The study of angles is one branch of **geometry**, which is concerned also with points, lines, surfaces and solids. Geometry probably originated in Ancient Egypt, where it was applied, for example, to land surveying and navigation. Indeed, the name 'geometry' is derived from the Greek for 'earth measure'.

It was Greek mathematicians, notably Euclid, who developed a theoretical foundation for geometry. He began his classic treatise *Elements* with definitions of basic geometrical concepts including angles, the subject of this chapter.

# 2.2 Angles

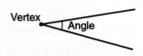

Vertex — Angle

**Figure 2.1**

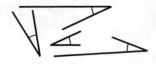

**Figure 2.2**

An **angle** is formed when two straight lines meet, as in Figure 2.1. The point where the lines meet is called the **vertex** of the angle. The size of the angle is the amount of turn from one line to the other, and is not affected by the lengths of the lines. The angles in Figure 2.2 are all equal.

If you begin by facing in a certain direction and then turn round until you are facing in the original direction again, as in Figure 2.3, you will have made a complete turn (or revolution). One complete turn is divided into 360 **degrees**, which is written as 360°.

**Figure 2.3**

In a half turn, Figure 2.4, therefore, there are 180°. In a quarter turn there are 90°, which is called a **right angle**. It is often marked on diagrams with a small square, as shown in Figure 2.5. If the angle between two lines is a right angle, the lines are said to be **perpendicular**.

Apart from angles of 90° and 180°, all other sizes of angles fall into one of three categories, shown in Figure 2.6.

An angle which is less than 90° is called an **acute** angle.

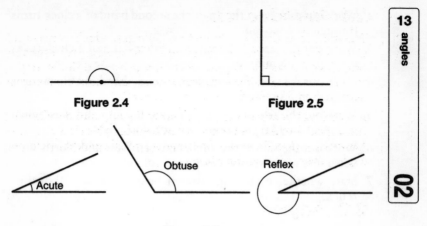

Figure 2.4

Figure 2.5

Figure 2.6

An angle which is between 90° and 180° is called an **obtuse** angle.

An angle which is greater than 180° is called a **reflex** angle.

There are three ways of naming angles.

One way is to use a letter inside the angle, as with the angle marked $x$ in Figure 2.7. Usually, a small letter, rather than a capital, is used.

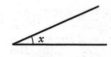

Figure 2.7

In Figure 2.8, the angle could be called angle $B$, but you should do this only if there is no risk of confusion. The symbols. $\angle$ and $\wedge$ are both used to represent 'angle', so angle $B$ could be shortened to $\angle B$ or $\widehat{B}$. It could be argued that there are two angles at $B$, one of them acute and one of them reflex, but you should assume that the acute angle is intended unless you are told otherwise.

Alternatively, the angle in Figure 2.8 could be called angle $ABC$ (or angle $CBA$). You could shorten this to $\angle ABC$ or $\widehat{ABC}$. The middle letter is always the vertex of the angle.

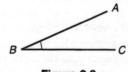

Figure 2.8

## EXERCISE 2.1

1 How many degrees does the hour hand of a clock turn in 20 minutes?

2 Work out the size of the angle the second hand of a clock turns through in 1 second.

3 Work out the size of the angle the hour hand of a clock turns through in 35 minutes.

4 Work out the size of the angle the second hand of a clock turns through in 55 seconds.

5 Work out the size of the angle between South and West when it is measured (a) clockwise, (b) anticlockwise.

6 Work out the size of the angle between West and North West when it is measured (a) clockwise, (b) anticlockwise.

7 State whether each angle drawn below is acute, obtuse or reflex.

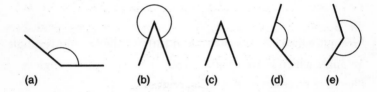

(a)     (b)     (c)     (d)     (e)

8 State whether each of the following angles is acute, obtuse or reflex.

(a) 172°     (b) 203°     (c) 302°     (d) 23°

9 Use capital letters to name the angles $x, y, z$ and $t$ in the figure.

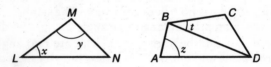

## 2.3 Measuring and drawing angles

It is a good idea to estimate the size of an angle before you try to measure it, as this will prevent you from giving answers which are not sensible. For this purpose, it is really only necessary to decide whether an angle is less than 90°, between 90° and 180°, between 180° and 270° or greater than 270°. With practice, you will probably be able to estimate an angle to within about 20°.

To estimate the size of the angle in Figure 2.9, notice that it is less than 90°. It is also more than half a right angle (45°), but nearer to 45° than 90°. A reasonable estimate is, therefore, 50° or 60°.

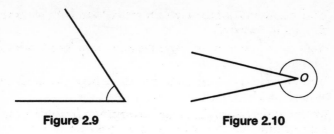

**Figure 2.9**                    **Figure 2.10**

The reflex angle at *O* in Figure 2.10 is between 270° and 360°. It is easier to estimate the size of the acute angle and then to subtract it from 360°.

The acute angle at *O* is about a quarter of a right angle, about 20° or 30°, so an estimate for the reflex angle at *O* is 330° or 340°.

To measure or draw an angle accurately, you use a **protractor**. The outer scale goes clockwise from 0° to 180° and the inner scale goes anticlockwise from 0° to 180°.

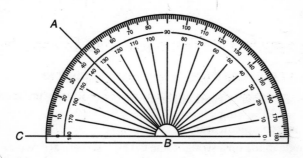

**Figure 2.11**

To measure the angle *ABC* in Figure 2.11, place the base line of the protractor on the line *BC* with the centre of the protractor over the vertex *B*. Read from the scale which has its zero on *BC*, that is, the outer scale. The size of angle *ABC* is 45°.

If you use the inner scale and give 135° as your measurement, you should see that this is not sensible, as angle *ABC* is acute.

## EXERCISE 2.2

1   Estimate the size of each of the angles overleaf and then measure them. Trace the angles and extend the lines so that you can use your protractor to measure the angles as accurately as possible.

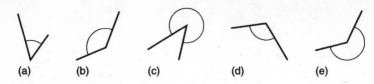

(a)    (b)    (c)    (d)    (e)

2 Draw angles with each of the following sizes.
   (a) 73°  (b) 246°  (c) 109°  (d) 281°  (e) 172°

## 2.4 Using angle facts

You should be able to measure and draw angles to within one or two degrees. Note, however, that many of the diagrams in the rest of this chapter are not drawn accurately but they are still helpful, even when you find solutions by carrying out a calculation.

This section deals with situations which use angle facts, most of which you met in Section 2.2.

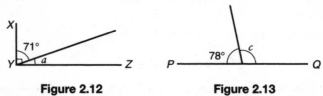

**Figure 2.12**            **Figure 2.13**

**There are 90° in a right angle.** In Figure 2.12, $\angle XYZ = 90°$, so $a + 71° = 90°$ giving $a = 90° - 71° = 19°$. Two angles whose sum is 90° are **complementary**.

The angle made on a straight line is 180°. If there are two or more angles on a straight line, then their sum must also be 180°. This fact is usually stated as **angles on a straight line add up to 180°**. In Figure 2.13, $PQ$ is a straight line, so $c$ and 78° add up to 180°. Thus, $c = 180° - 78° = 102°$. Two angles which add up to 180° are called **supplementary**, so 78° and 102° are supplementary angles.

There are 360° in a complete turn. The angles in Figure 2.14, when added together, make a complete turn. They are called **angles at a point**. So **the angles at a point add up to 360°**. Therefore 117°, 122° and $e$ add to 360°. $117° + 122° = 239°$, $e = 360° - 239° = 121°$.

When two straight lines cross, as in Figure 2.15, four angles, $a, b, c$ and $d$, are formed. Since $a$ added to $b$ makes 180°, and $c$ added to $b$ also makes 180°, it follows that $a = c$. Similarly, $b = d$.

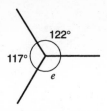

Figure 2.14

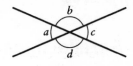

Figure 2.15

The angles *a* and *c* are called **vertically opposite**. The angles *b* and *d* are also vertically opposite. The argument in the previous paragraph shows that **vertically opposite angles are equal**.

## Summary

There are 90° in a right angle.

The angles on a straight line add up to 180°.

The angles at a point add up to 360°.

Vertically opposite angles are equal.

### EXERCISE 2.3

1 Find the size of each angle marked with a letter. Where they occur, *AB* and *CD* are straight lines.

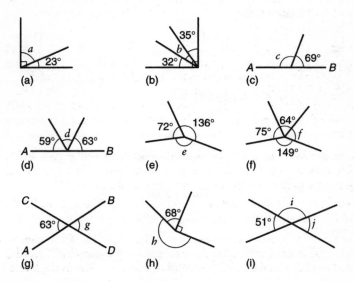

2 In the diagram, is ∠AOD a straight line? Give a reason for your answer.

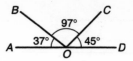

## 2.5 Parallel lines

The word 'parallel', derived from the Greek, means 'alongside one another'. Straight lines are **parallel** if they are always the same distance apart. Parallel lines never meet, no matter how far they are extended. Straight railway lines, for example, are parallel.

**Figure 2.16**

**Figure 2.17**

In diagrams, arrows are often used to show that lines are parallel. (See Figure 2.16.) A straight line which crosses two or more parallel lines is called a **transversal**. Figure 2.17 shows a transversal.

**Figure 2.18**

**Figure 2.19**

In Figure 2.18, a transversal cuts a pair of parallel lines. The pair of shaded angles are called **corresponding angles** and are equal to each other. You will see from Figure 2.19 that the corresponding angles form an 'F-shape'. Looking for such a shape may help you to find pairs of corresponding angles.

**Figure 2.20**

Other pairs of corresponding angles have been shaded in Figure 2.20.

In Figure 2.21, a transversal cuts a pair of parallel lines. The pair of shaded angles are **alternate angles,** and are equal to each other.

**Figure 2.21**          **Figure 2.22**          **Figure 2.23**

The characteristic 'Z-shape' in Figure 2.22 can help you spot alternate angles. Another pair of alternate angles has been shaded in Figure 2.23.

In Figure 2.24, the angles marked *a* and *b* are equal because they are corresponding, and the angles marked *b* and *c* are equal because they are vertically opposite. So the alternate angles *a* and *c* are equal.

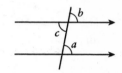

**Figure 2.24**

---

**Example 2.5.1**

Find, with reasons, the sizes of the angles marked *a* and *b*.

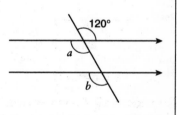

Angles *a* and 120° are vertically opposite, so $a = 120°$.

Angles *a* and *b* are corresponding, so $b = 120°$.

---

### EXERCISE 2.4

1 Write down the letter of the angle which is (i) corresponding to the shaded angle, (ii) alternate to the shaded angle.

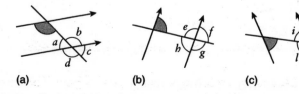

(a)          (b)          (c)

2 Find the size of each of the marked angles.

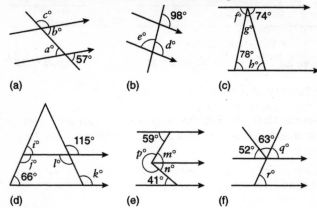

(a)

(b)

(c)

(d)

(e)

(f)

## 2.6 Bearings

In some situations, compass bearings, based on North, East, South and West, are used to describe directions. For navigation, **three-figure bearings** are used. A three-figure bearing is the angle always measured clockwise from North. When the angle is less than 100°, one or two zeros are inserted in front of the angle, so that the bearing still has three figures. Thus, East as a bearing is 090° and South is 180°. The directions of the four bearings 032°, 143°, 227° and 315° are shown in Figure 2.25.

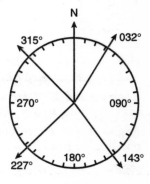

Figure 2.25

### Example 2.6.1

The bearing of $B$ from $A$ is 074°. Find the bearing of $A$ from $B$.

It is important to draw a rough sketch.

Start by putting *A* on the paper with the line due North through *A*, and then put *B* on the diagram, together with the line due North through *B*.

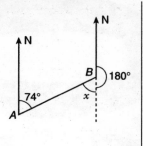

The bearing of *A* from *B* is the reflex angle at *B* and is therefore $180° + x$. But the angle $x$ is alternate to the angle of $74°$, so $x = 74°$.

Therefore the bearing of *A* from *B* is $180° + x = 254°$.

## EXERCISE 2.5

**1** Make a tracing of the figure and extend the lines. Then, using your protractor, find the bearings of *A*, *B*, *C* and *D* from *O*.

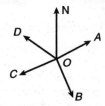

**2** Draw accurate diagrams to illustrate these bearings.

(a) 83°    (b) 126°    (c) 229°    (d) 284°

**3** Express the following directions as three-figure bearings.

(a) West         (b) North east
(c) South west    (d) North west

**4** Express the following directions as three-figure bearings.

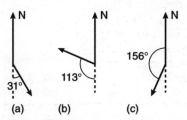

**5** The bearing of *B* from *A* is 139°. Find the bearing of *A* from *B*.

**6** The bearing of *D* from *C* is 205°. Find the bearing of *C* from *D*.

**03**

**fractions**

**In this chapter you will learn:**

- about equivalent fractions
- how to compare fractions
- how to add and subtract fractions
- how to multiply and divide fractions.

# 3.1 Introduction

The word 'fraction' comes from the Latin *fractus* (broken) and fractions were often called 'broken numbers'. For many centuries, the numbers 1, 2, 3, 4 etc. met all man's needs and, although the Babylonians developed a system of fractions in about 2000 BC, it was 400 years later that the Ancient Egyptians produced the first thorough treatment.

# 3.2 What is a fraction?

A fraction is one or more equal parts of a whole.

In Figure 3.1, the left circle has been split into 4 equal parts (quarters) and 3 of them are shaded. The **fraction** of the circle which is shaded is $\frac{3}{4}$. The top number, 3, is called the **numerator** and the bottom number, 4, is called the **denominator.** One of the four parts is unshaded, so the fraction of the circle which is unshaded is $\frac{1}{4}$.

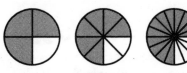

**Figure 3.1**

If you were asked what fraction of the circle is shaded in the middle circle in Figure 3.1, you might give $\frac{6}{8}$ as your answer, as the circle has been split into 8 equal parts and 6 of them are shaded. Similarly, in the third circle, the fraction $\frac{12}{16}$ is shaded.

You might say that it is obvious from Figure 3.1 that the three fractions $\frac{3}{4}$, $\frac{6}{8}$ and $\frac{12}{16}$ are equal. You would be right: they are equal in value and they are called **equivalent fractions.** You can write $\frac{3}{4} = \frac{6}{8} = \frac{12}{16}$.

You could get $\frac{6}{8}$ from $\frac{3}{4}$ by multiplying both the numerator and the denominator of $\frac{3}{4}$ by 2, that is, $\frac{3 \times 2}{4 \times 2} = \frac{6}{8}$. You can get $\frac{12}{16}$ by multiplying the numerator and the denominator of $\frac{3}{4}$ by 4. Similarly, $\frac{9}{12}$, $\frac{15}{20}$ and $\frac{18}{24}$ are three more fractions equivalent to $\frac{3}{4}$. In general, if you multiply the numerator and the denominator of a fraction by the same number, you will obtain a fraction equivalent to the original.

## EXERCISE 3.1

1  What fractions of these shapes are (a) shaded, (b) unshaded?

2  If $\frac{5}{6}$ of a shape is shaded, what fraction is unshaded?

3  If $\frac{3}{7}$ of a shape is shaded, what fraction is unshaded?

4  From $\frac{4}{3}, \frac{3}{8}, \frac{5}{6}$ and $\frac{8}{9}$, write down a fraction with

   (a) a numerator of 5,    (b) a denominator of 8.

5  Write down five fractions which are equivalent to

   (a) $\frac{1}{3}$   (b) $\frac{3}{4}$   (c) $\frac{2}{5}$   (d) $\frac{5}{9}$

6  Complete $\frac{2}{3} = \frac{}{24}$ and $\frac{5}{8} = \frac{}{24}$. Is $\frac{2}{3}$ larger than $\frac{5}{8}$?

## 3.3  Which fraction is bigger?

If two fractions have equal denominators, it is easy to see which fraction is the larger. For example, $\frac{5}{8}$ is clearly larger than $\frac{3}{8}$. You only have to compare their numerators and see that 5 is larger than 3.

To find which is the larger of $\frac{2}{3}$ and $\frac{3}{4}$ make lists of fractions equivalent to $\frac{2}{3}$ and to $\frac{3}{4}$. Such lists are $\frac{2}{3} = \frac{4}{6} = \frac{6}{9} = \frac{8}{12} = \frac{10}{15} = \ldots$ and $\frac{3}{4} = \frac{6}{8} = \frac{9}{12} = \frac{12}{16} = \frac{15}{20} = \ldots$. In each list there is a fraction with a denominator of 12 : $\frac{2}{3} = \frac{8}{12}$ and $\frac{3}{4} = \frac{9}{12}$. So $\frac{3}{4}$ is larger than $\frac{2}{3}$.

A quicker method is to convert $\frac{2}{3}$ and $\frac{3}{4}$ to equivalent fractions with the same denominator, in this case 12, because 12 is the lowest number which is a multiple of both 3 and 4. This number, 12, is called the **lowest common multiple** of 3 and 4.

## 3.4  Simplifying fractions

The fractions $\frac{2}{3}$ and $\frac{8}{12}$ are equivalent. To show this, start with $\frac{2}{3}$ and multiply both the numerator and the denominator by 4.

Alternatively, start with $\frac{8}{12}$ and divide both its numerator and its denominator by 4. In general, by dividing both the numerator and the denominator of a fraction by the same number, you obtain a fraction which is equivalent to the original fraction. This process is called **cancelling**.

If the numerator and denominator of a fraction have no common factors, the fraction is said to be in its simplest form or in its **lowest terms**. So $\frac{2}{3}$ is **the simplest** form of $\frac{16}{24}$ but $\frac{4}{10}$ is not the simplest form of $\frac{12}{30}$, because 2 is a factor of both 4 and of 10. If you divide the numerator and the denominator of $\frac{4}{10}$ by 2, you obtain $\frac{2}{5}$, which is in its simplest form.

To find the simplest form of $\frac{12}{30}$ in a single step, you must find the highest number which is a factor of both 12 and 30, which is 6, and then divide both the numerator and the denominator by 6.

### EXERCISE 3.2

1 Which fraction is larger?

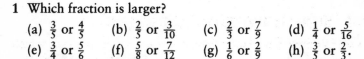

    (a) $\frac{3}{5}$ or $\frac{4}{5}$     (b) $\frac{2}{5}$ or $\frac{3}{10}$     (c) $\frac{2}{3}$ or $\frac{7}{9}$     (d) $\frac{1}{4}$ or $\frac{5}{16}$

    (e) $\frac{3}{4}$ or $\frac{5}{6}$     (f) $\frac{5}{8}$ or $\frac{7}{12}$     (g) $\frac{1}{6}$ or $\frac{2}{9}$     (h) $\frac{3}{5}$ or $\frac{2}{3}$.

2 Put these fractions in order of size starting with the smallest.

    (a) $\frac{1}{2}, \frac{5}{8}, \frac{7}{12}$     (b) $\frac{3}{4}, \frac{4}{5}, \frac{7}{10}$

3 Complete each of the following.

    (a) $\frac{4}{8} = \frac{}{4}$     (b) $\frac{6}{18} = \frac{}{6}$     (c) $\frac{20}{24} = \frac{}{6}$

4 Complete each of the following.

    (a) $\frac{30}{40} = \frac{6}{}$     (b) $\frac{24}{36} = \frac{4}{}$     (c) $\frac{18}{27} = \frac{2}{}$

5 Find the simplest form of each of the following.

    (a) $\frac{4}{6}$    (b) $\frac{9}{12}$    (c) $\frac{5}{20}$    (d) $\frac{8}{10}$

    (e) $\frac{18}{21}$    (f) $\frac{24}{45}$    (g) $\frac{7}{28}$    (h) $\frac{33}{55}$

6 Find the simplest form of each of the following.

    (a) $\frac{4}{12}$    (b) $\frac{10}{20}$    (c) $\frac{6}{18}$    (d) $\frac{24}{40}$

    (e) $\frac{45}{54}$    (f) $\frac{36}{60}$    (g) $\frac{40}{100}$    (h) $\frac{45}{75}$

## 3.5 Improper fractions

Figure 3.2 shows that there are 6 thirds in 2 wholes, that is, $\frac{6}{3} = 2$. Notice also that $\frac{6}{3}$ simplifies to $\frac{2}{1}$, so $\frac{2}{1} = 2$. You can think of the fraction line as a division sign, so $6 \div 3 = \frac{6}{3}$. Hence $6 \div 3 = \frac{6}{3} = \frac{2}{1} = 2$.

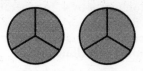

**Figure 3.2**

A fraction such as $\frac{6}{3}$, in which the numerator is greater than the denominator, is called an **improper fraction**. All the fractions you have met so far in this chapter have been **proper** fractions, in which the numerator is less than the denominator.

If the denominator divides exactly into the numerator the improper fraction is equal to a whole number, but this doesn't always happen.

Figure 3.3 shows that there are 7 quarters in $1\frac{3}{4}$, that is, $\frac{7}{4} = 1\frac{3}{4}$. The number $1\frac{3}{4}$ is called a **mixed number**, because it is a mixture of a whole number and a proper fraction.

To change $\frac{7}{4}$ to a mixed number, think of it as $7 \div 4$. Since 4 divides into 7 once with a remainder of 3, the whole number is 1 and the numerator of the fraction is 3. The denominator, 4, of the fraction is the same as in the improper fraction.

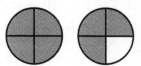

**Figure 3.3**

---

**Example 3.5.1**

Change $\frac{14}{3}$ to a mixed number.

Think of $\frac{14}{3}$ as $14 \div 3$. Then divide 14 by 3 to get 4 with a remainder of 2. Hence $\frac{14}{3} = 14 \div 3 = 4\frac{2}{3}$.

If you want to change an improper fraction which is not in its simplest form to a mixed number, express it in its simplest form and then change it to a mixed number.

To change a mixed number such as $3\frac{2}{5}$ to an improper fraction, you need to work out how many fifths there are in $3\frac{2}{5}$. There are 5 fifths in 1, so there are 15, that is $3 \times 5$, fifths in 3. Adding the extra 2 fifths gives 17 fifths, which is $\frac{17}{5}$. So $3\frac{2}{5} = \frac{17}{5}$.

You would write down this process as $3\frac{2}{5} = \frac{(5\times3)+2}{5} = \frac{17}{5}$.

## EXERCISE 3.3

1 Write each of the following as a whole number.
   (a) $\frac{30}{6}$    (b) $\frac{70}{10}$    (c) $\frac{42}{7}$    (d) $\frac{56}{8}$

2 Write the following as improper fractions in their simplest forms.
   (a) 3       (b) 12

3 Change the following to mixed numbers.
   (a) $13 \div 5$    (b) $19 \div 4$    (c) $25 \div 6$    (d) $15 \div 8$
   (e) $\frac{20}{3}$       (f) $\frac{18}{5}$       (g) $\frac{17}{8}$       (h) $\frac{29}{9}$

4 Change the following to mixed numbers in their simplest forms.
   (a) $\frac{14}{4}$    (b) $\frac{21}{6}$    (c) $\frac{25}{15}$    (d) $\frac{35}{14}$

5 Change the following to improper fractions.
   (a) $2\frac{4}{5}$    (b) $4\frac{2}{3}$    (c) $3\frac{1}{2}$    (d) $5\frac{3}{4}$

# 3.6 Adding and subtracting fractions

Just as it is easy to order fractions with equal denominators, it is easy to add and subtract them.

Figure 3.4 shows that $\frac{3}{7} + \frac{2}{7} = \frac{5}{7}$, that is, you add the numerators but you do not add the denominators. $\frac{3}{7} + \frac{2}{7}$ is not equal to $\frac{3}{7} + \frac{2}{7} = \frac{5}{14}$. The fraction $\frac{5}{7}$ is in its simplest form but sometimes you will be able to simplify your answer. You should do so if you can.

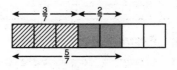

Figure 3.4

You can also subtract fractions with equal denominators. For example, $\frac{7}{9} - \frac{2}{9} = \frac{5}{9}$, that is, you subtract the second numerator from the first but, as with addition, the denominator stays the same.

If fractions have different denominators, convert them to equivalent fractions with equal denominators. To add $\frac{3}{5}$ and $\frac{1}{4}$, for example, convert both to equivalent fractions with denominator 20, the lowest common multiple of the denominators 5 and 4. (See Section 3.3.)

Then $\frac{3}{5} + \frac{1}{4} = \frac{12}{20} + \frac{5}{20} = \frac{17}{20}$.

---

**Example 3.6.1**

Work out $5\frac{2}{3} + 2\frac{1}{2}$.

Adding the whole numbers, $5 + 2 = 7$.

The lowest common multiple of the denominators 2 and 3 is 6.

Then $\frac{2}{3} + \frac{1}{2} = \frac{4}{6} + \frac{3}{6} = \frac{4+3}{6} = \frac{7}{6}$, and, as a mixed number $\frac{7}{6} = 1\frac{1}{6}$.

Finally, adding the two sums $7 + 1\frac{1}{6} = 8\frac{1}{6}$.

---

You can use the same method for *some* subtractions.

---

**Example 3.6.2**

Work out $4\frac{2}{3} - 1\frac{3}{5}$.

Subtracting the whole numbers, $4 - 1 = 3$.

The lowest common multiple of the denominators is 15.

Then $\frac{2}{3} - \frac{3}{5} = \frac{10}{15} - \frac{9}{15} = \frac{10-9}{15} = \frac{1}{15}$,
$$\text{and}$$
$$4\frac{2}{3} - 1\frac{3}{5} = 3 + \frac{1}{15} = 3\frac{1}{15}.$$

---

If you try to use this method with $4\frac{3}{5} - 1\frac{2}{3}$, the subtraction $\frac{9}{15} - \frac{10}{15}$ arises, that is, you have to subtract a larger fraction from a smaller. There are ways around this difficulty, but the safest is to start by changing both mixed numbers to improper fractions. The disadvantage of this method is that the numerators can get large.

**Example 3.6.3**

Work out $4\frac{3}{5} - 1\frac{2}{3}$.

As improper fractions $4\frac{3}{5} = \frac{23}{5}$ and $1\frac{2}{3} = \frac{5}{3}$.

The lowest common multiple of the denominators is 15.

Then $\frac{23}{5} - \frac{5}{3} = \frac{69}{15} - \frac{25}{15} = \frac{69-25}{15} = \frac{44}{15} = 2\frac{14}{15}$.

## EXERCISE 3.4

Work out the following. Give every answer in its simplest form.

1 (a) $\frac{3}{5} + \frac{1}{5}$    (b) $\frac{4}{7} + \frac{2}{7}$    (c) $\frac{3}{10} + \frac{1}{10}$    (d) $\frac{7}{12} + \frac{1}{12}$

   (e) $\frac{7}{9} + \frac{4}{9}$    (f) $\frac{7}{8} + \frac{3}{8}$    (g) $\frac{3}{10} + \frac{7}{10}$    (h) $\frac{9}{10} + \frac{7}{10}$

2 (a) $\frac{4}{5} - \frac{1}{5}$    (b) $\frac{5}{7} - \frac{3}{7}$    (c) $\frac{5}{8} - \frac{3}{8}$    (d) $\frac{5}{6} - \frac{1}{6}$

3 (a) $\frac{1}{3} + \frac{2}{5}$    (b) $\frac{3}{8} + \frac{1}{4}$    (c) $\frac{1}{6} + \frac{1}{2}$    (d) $\frac{2}{3} + \frac{5}{6}$

   (e) $\frac{2}{3} + \frac{3}{5}$    (f) $\frac{1}{6} + \frac{4}{9}$    (g) $\frac{7}{8} + \frac{5}{12}$    (h) $\frac{5}{6} + \frac{1}{2}$

4 (a) $2\frac{1}{4} + 3\frac{4}{5}$    (b) $4\frac{3}{5} + 1\frac{3}{10}$    (c) $5\frac{1}{2} + 2\frac{3}{10}$    (d) $1\frac{5}{6} + 3\frac{3}{4}$

5 (a) $3\frac{7}{8} - 1\frac{1}{4}$    (b) $4\frac{2}{3} - 2\frac{1}{2}$    (c) $3\frac{5}{6} - 2\frac{3}{4}$    (d) $4\frac{2}{3} - 1\frac{5}{6}$

# 3.7 Multiplication of fractions

Multiplication by a whole number is the same as repeated addition. Thus, $3 \times 2 = 3 + 3$ and $\frac{3}{7} \times 2 = \frac{3}{7} + \frac{3}{7} = \frac{6}{7}$. The numerator, 3, has been multiplied by the whole number, 2, but the denominator is unchanged. A common error is to multiply *both* the numerator and the denominator by 2 but this gives a fraction, $\frac{6}{14}$, which is *equivalent* to $\frac{3}{7}$.

**Example 3.7.1**

Work out $\frac{5}{6} \times 8$.

Multiplying the numerator by the whole number, $\frac{5}{6} \times 8 = \frac{40}{6}$.

As a mixed number $\frac{40}{6} = 6\frac{4}{6}$, which, in its simplest form, is $6\frac{2}{3}$.

The multiplication $\frac{4}{5} \times \frac{2}{3}$ is illustrated in Figure 3.5, where you can see that $\frac{8}{15}$ of the total square is shaded. Thus $\frac{4}{5} \times \frac{2}{3} = \frac{8}{15}$. This suggests that to multiply two fractions you multiply the numerators and then you multiply the denominators.

**Figure 3.5**

Multiplying a fraction by a whole number is similar to multiplying one fraction by another fraction. Writing the whole number as an improper fraction with the whole number as the numerator and 1 as the denominator, enables you to write, for example, 10 as $\frac{10}{1}$. Then $\frac{3}{4} \times 10 = \frac{3}{4} \times \frac{10}{1} = \frac{30}{4} = \frac{15}{2} = 7\frac{1}{2}$.

To multiply two mixed numbers, change them to improper fractions.

---

**Example 3.7.2**

Work out $2\frac{1}{4} \times 1\frac{2}{5}$.

$2\frac{1}{4} \times 1\frac{2}{5} = \frac{9}{4} \times \frac{7}{5} = \frac{9 \times 7}{4 \times 5} = \frac{63}{20} = 3\frac{3}{20}$.

---

## 3.8 Fractions of quantities

One half of 10 is obviously 5. Therefore $\frac{1}{2}$ of 10 is 5.

But $\frac{1}{2} \times 10 = \frac{1}{2} \times \frac{10}{1} = \frac{1 \times 10}{2 \times 1} = \frac{10}{2} = 5$.

This suggests that, in mathematics, '**of**' means multiply. You can use this to find fractions of physical quantities.

---

**Example 3.8.1**

Work out $\frac{2}{3}$ of 27 miles.

$\frac{2}{3}$ of $27 = \frac{2}{3} \times 27 = \frac{2}{3} \times \frac{27}{1} = \frac{54}{3} = 18$, so $\frac{2}{3}$ of 27 miles is 18 miles.

---

## EXERCISE 3.5

Work out the following. Give every answer in its simplest form.

1  (a) $\frac{2}{3} \times 5$     (b) $\frac{4}{3} \times 15$     (c) $\frac{3}{4} \times 10$     (d) $40 \times \frac{7}{8}$

2  (a) $\frac{3}{4} \times \frac{5}{7}$     (b) $\frac{2}{3} \times \frac{7}{8}$     (c) $\frac{5}{6} \times \frac{9}{10}$     (d) $\frac{3}{5} \times \frac{2}{9}$

3  (a) $2\frac{1}{2} \times 1\frac{3}{4}$     (b) $1\frac{7}{8} \times 2\frac{4}{5}$     (c) $2\frac{2}{5} \times 3\frac{1}{3}$     (d) $\frac{5}{8} \times 2\frac{1}{3}$

   (e) $4\frac{1}{2} \times \frac{4}{9}$     (f) $\frac{5}{6} \times 1\frac{1}{2}$     (g) $1\frac{2}{3} \times \frac{3}{7}$     (h) $4 \times 3\frac{1}{2}$

4  (a) $\frac{1}{6}$ of 30     (b) $\frac{5}{8}$ of 56     (c) $\frac{3}{5}$ of 45     (d) $\frac{3}{4}$ of 44

5  (a) $\frac{2}{3}$ of 48 kg     (b) $\frac{3}{10}$ of 70 cm     (c) $\frac{4}{5}$ of 60 litres

## 3.9  Division of fractions

The calculation $12 \div 3$ means 'what do you multiply three by to get 12?' Similarly, $5 \div \frac{1}{2}$ means 'what do you multiply one half by to get 5?' There are 2 halves in 1; so there are $5 \times 2$, that is 10, halves in 5.

By first writing the whole number as an improper fraction, $5 \div \frac{1}{2} = \frac{5}{1} \div \frac{1}{2} = 10$.

However, you also know that $5 \times 2 = \frac{5}{1} \times \frac{2}{1} = \frac{10}{1} = 10$. This suggests that dividing by $\frac{1}{2}$ is the same as multiplying by 2.

Now think about $6 \div \frac{3}{4}$. This means 'how many times do you multiply $\frac{3}{4}$ to get 6 ?' First notice that $\frac{4}{3} \times \frac{3}{4} = \frac{12}{12} = 1$, so you multiply $\frac{3}{4}$ by $\frac{4}{3}$ to get 1, so you must multiply $\frac{3}{4}$ by $6 \times \frac{4}{3}$ to get 6. Therefore

$$6 \div \frac{3}{4} = 6 \times \frac{4}{3} = \frac{6}{1} \times \frac{4}{3} = \frac{24}{3} = 8.$$

This suggests a way to divide one fraction by another. You turn the second fraction turn upside down and multiply the first fraction by it.

---

**Example 3.9.1**

Work out    (a) $\frac{2}{3} \div \frac{3}{4}$     (b) $1 \div \frac{4}{5}$.

(a) $\frac{2}{3} \div \frac{3}{4} = \frac{2}{3} \times \frac{4}{3} = \frac{8}{9}$.

(b) $1 \div \frac{4}{5} = \frac{1}{1} \times \frac{5}{4} = \frac{5}{4} = 1\frac{1}{4}$.

---

The result of dividing a number into 1 is called the **reciprocal** of the number. So $\frac{5}{4}$ is the reciprocal of $\frac{4}{5}$. You can also work out that $\frac{4}{5}$ is the reciprocal of $\frac{5}{4}$ so $\frac{5}{4}$ and $\frac{4}{5}$ are reciprocals of each other.

Note that, as with multiplication, mixed numbers must always be changed to improper fractions before dividing.

---

**Example 3.9.2**

Work out $3\frac{1}{2} \div \frac{7}{12}$.

$3\frac{1}{2} \div \frac{7}{12} = \frac{7}{2} \times \frac{12}{7} = \frac{84}{14} = 6.$

---

### EXERCISE 3.6

Work out the following. Give every answer in its simplest form.

1 (a) $2 \div \frac{1}{3}$     (b) $4 \div \frac{3}{4}$     (c) $6 \div \frac{2}{3}$     (d) $2 \div \frac{3}{5}$

2 (a) $\frac{4}{5} \div \frac{3}{8}$     (b) $\frac{5}{12} \div \frac{3}{4}$     (c) $\frac{7}{9} \div \frac{2}{3}$     (d) $\frac{3}{5} \div \frac{8}{9}$

   (e) $\frac{2}{5} \div 3$     (f) $\frac{7}{10} \div \frac{4}{5}$     (g) $\frac{11}{12} \div \frac{3}{8}$     (h) $\frac{4}{9} \div \frac{7}{12}$

3 (a) $2\frac{1}{2} \div 1\frac{7}{8}$     (b) $2\frac{3}{4} \div 3\frac{1}{3}$     (c) $2\frac{5}{8} \div \frac{7}{16}$     (d) $3 \div 4\frac{1}{2}$

   (e) $2\frac{7}{10} \div 1\frac{1}{3}$     (f) $1\frac{3}{4} \div 2\frac{1}{4}$     (g) $\frac{3}{8} \div 1\frac{4}{5}$     (h) $2\frac{3}{5} \div \frac{2}{3}$

## 3.10  A number as a fraction of another number

Fractions are a useful way of comparing two quantities. For example, 90° is often called a $\frac{1}{4}$-turn, because there are 360° in a complete turn and $\frac{90}{360} = \frac{1}{4}$ in its simplest form. Similarly, 45 minutes is $\frac{3}{4}$ hour, as there are 60 minutes in an hour and $\frac{45}{60} = \frac{3}{4}$ in its simplest form. The numerator and the denominator must be in the same units.

---

**Example 3.10.1**

Express, in its simplest form, 36 minutes as a fraction of 2 hours.

There are 120 minutes in 2 hours. The required fraction is $\frac{36}{120} = \frac{3}{10}$.

---

# EXERCISE 3.7

Express the first quantity as a fraction of the second. Give each answer in its simplest form.

1  (a)  8 minutes, 1 hour    (b)  15 hours, 1 day
   (c)  5 days, 2 weeks      (d)  48 seconds, 2 minutes
   (e)  50 minutes, 2 hours  (f)  6 days, 3 weeks
   (g)  60 cm, 1 metre       (h)  18 hours, 4 days
   (i)  70 cm, 2 metres      (j)  14 hours, 1 week

2  On a jury there are 8 women and 4 men. What fraction of the jury are women?

3  Find a fraction which is halfway between $\frac{3}{5}$ and $\frac{4}{5}$.

4  Two fifths of the employees in a company drive to work, one third travel by bus and the rest walk. Find the fraction who walk.

5  Last Friday, Tony worked for $7\frac{1}{2}$ hours. Express this time as a fraction of a day.

6  The sum of two mixed numbers is $6\frac{3}{20}$. One of the mixed numbers is $2\frac{3}{4}$. Work out the other one.

7  The product of two fractions is $\frac{7}{15}$. One of the fractions is $\frac{4}{5}$. Work out the other one.

8  An angle is $\frac{2}{3}$ of a turn. Express this in degrees.

9  When a petrol tank is $\frac{7}{12}$ full, it contains $5\frac{1}{4}$ gallons. How many gallons does it hold when full?

# 04

## two-dimensional shapes

**In this chapter you will learn:**

- about different types of triangles and polygons
- how to use the angle sums of triangles and polygons
- how to construct triangles
- about the symmetries of triangles and polygons
- about tessellations.

# 4.1 Introduction

Two-dimensional shapes are **flat** shapes, such as squares and circles. They are also called **plane** shapes; the geometry which deals with them is called **plane geometry**. Thales of Miletus, a Greek, laid the foundations of plane geometry in the sixth century BC. About 200 years later, Euclid's *Elements* contained a complete, rigorous treatment of it.

# 4.2 Triangles

The simplest two-dimensional shape is the **triangle,** a three-sided shape with three angles (Figure 4.1). As a triangle made of rods cannot be deformed, it is widely used in buildings and other structures.

**Figure 4.1**            **Figure 4.2**            **Figure 4.3**

Draw a triangle on paper and label its angles *a*, *b* and *c*. See Figure 4.2. Tear off the corners and fit angles *a*, *b* and *c* together, as in Figure 4.3. They fit approximately along a straight line. From Chapter 02, the three angles on a straight line add up to 180°, so $a + b + c = 180°$. This suggests that the angle sum of a triangle is 180°.

However, this demonstration is approximate. It only works for the triangle that you drew. It does not show that the sum of the angles in a triangle is 180° for any triangle. To do that, you need a **proof**.

In Figure 4.4, the side *AB* of triangle *ABC* has been extended and, at *B*, a line has been drawn parallel to *AC*. Angles *f*, *g* and *h* lie along a straight line, so $f + g + h = 180°$. (See Section 2.5.)

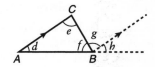

**Figure 4.4**

As angles *g* and *e* are alternate, $g = e$. As angles *h* and *d* are corresponding, $h = d$. So $f + e + d = 180°$.

Therefore **the angle sum of a triangle is 180°**.

The triangle in Figure 4.5 is **acute-angled,** as its angles are all less than 90°. An **obtuse-angled triangle** (Figure 4.6) has an obtuse angle and a **right-angled triangle** (Figure 4.7) has a right angle.

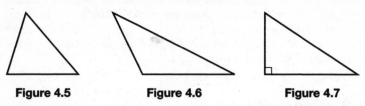

**Figure 4.5**          **Figure 4.6**          **Figure 4.7**

There are three other types of triangle. An **equilateral triangle** has three equal sides and angles. As the angle sum is 180°, each angle is $180° \div 3 = 60°$. In Figure 4.8, the marked sides show that they are equal.

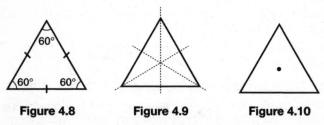

**Figure 4.8**          **Figure 4.9**          **Figure 4.10**

An equilateral triangle has 3 **lines of symmetry,** shown dotted in Figure 4.9. If the triangle is folded along one of its lines of symmetry, one half of the triangle will fit exactly on top of the other half. An alternative name for a line of symmetry is a **mirror line,** as one half of the triangle is a mirror image, or **reflection,** of the other half.

An equilateral triangle has **rotational symmetry** of **order** 3. This is the number of ways a tracing of the shape will fit on top of itself without turning it over. In Figure 4.10, the point of the triangle about which the tracing paper turns is called the **centre of rotational symmetry**.

A triangle which has two equal sides is called an **isosceles** triangle. In an isosceles triangle the angles opposite the equal sides are also equal (see Figure 4.11). The reverse is also true: if two angles of a triangle are equal, then the sides opposite them are also equal, so the triangle is isosceles. 'Isosceles' is a Greek word, meaning literally 'equal legs'.

An isosceles triangle generally has one line of symmetry (Figure 4.12) and rotational symmetry of order 1. That is, a tracing of the triangle will fit on top of itself in only one way.

In the special case when the isosceles triangle is equilateral, it has three lines of symmetry.

 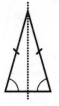 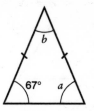

| **Figure 4.11** | **Figure 4.12** | **Figure 4.13** |

---

**Example 4.2.1**

In Figure 4.13, find the sizes of the angles marked *a* and *b*.

The triangle is isosceles, so the angles opposite the equal sides are equal, and $a = 67°$.

As the angle sum of the triangle is 180°,
$b = 180° - 2 \times 67° = 46°$.

---

Triangles with no equal sides and no equal angles are **scalene** triangles. They have no lines of symmetry and rotational symmetry of order 1.

## EXERCISE 4.1

In Questions 1–9, find the size of each of the angles marked with letters. In some questions you will need to use facts about angles from Chapter 02. The triangles are not drawn to scale.

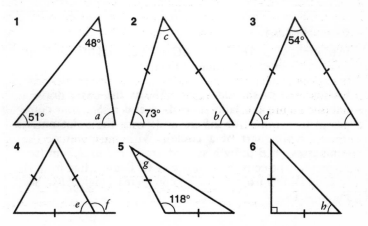

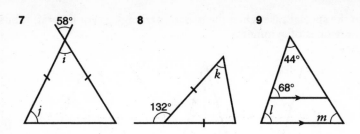

7  8  9

10  One of the angles of an isosceles triangle is 74°. The other two angles are $x$ and $y$. Find two possible pairs of values for $x$ and $y$.

11  One of the equal sides of an isosceles triangle is extended, and the angle between the extended side and the other equal side is 40°. Find the angles in the triangle.

## 4.3  Constructing triangles

Given three suitable measurements, you can make an accurate drawing of a triangle with a ruler, a pair of compasses and a protractor. If the corners, or vertices, of the triangle are labelled $A$, $B$ and $C$, as in Figure 4.14, the length of the side joining $A$ and $B$ is denoted by $AB$.

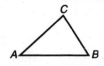

**Figure 4.14**

---

**Example 4.3.1**

In triangle $ABC$, $AB = 3$ cm, $AC = 2.5$ cm and $BC = 2.7$ cm. Make an accurate drawing of the triangle.

Choose one of the sides, say $AB$, as the base, draw a line 3 cm long and label its ends $A$ and $B$. Set your compasses to a radius of 2.5 cm and, with the point on $A$, draw an arc (part of a circle). With the point of the compasses on $B$ draw a second arc of radius 2.7 cm (see Figure 4.15) to cut the first arc at $C$. Draw a line from $A$ to $C$ and a line from $B$ to $C$ to complete triangle $ABC$. See Figure 4.16.

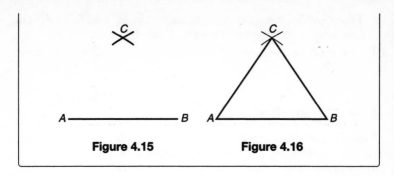

**Figure 4.15**　　　　　**Figure 4.16**

You can construct a triangle if you are given the length of one side and the sizes of two angles. If necessary, you can, of course, work out the size of the third angle of the triangle.

You can also construct a triangle if you are given the lengths of two sides and the size of the angle between the two given sides.

If you are given two sides and an angle other than the included angle, you can sometimes construct two different triangles from the information. (See Exercise 4.2, Question 9.)

If you are given three angles whose sum is 180° and no side, you can construct an infinite number of triangles from the information.

### EXERCISE 4.2

In Questions 1–7, make an accurate drawing of the triangle $ABC$.

1 $AB = 4.2$ cm, $AC = 3.7$ cm, $BC = 3.1$ cm.
2 $AB = 3.6$ cm, $AC = 2.8$ cm, $BC = 5.2$ cm.
3 $AB = 4.3$ cm, angle $A = 59°$, angle $B = 42°$.
4 $AB = 4.8$ cm, angle $A = 124°$, angle $B = 21°$.
5 $AB = 4.3$ cm, angle $A = 25°$, angle $C = 27°$.
6 $AB = 5.1$ cm, $AC = 4.8$ cm, angle, $A = 43°$.
7 $AB = 4.5$ cm, $BC = 3.9$ cm, angle $B = 136°$.
8 What happens if you try to construct a triangle $ABC$ with $AB = 6$ cm, $AC = 3$ cm and $BC = 2$ cm? Find a rule which tells you whether, given three lengths, it is possible to construct a triangle with sides of these lengths.

9  Triangle $ABC$ has $AB = 5.7$ cm, angle $A = 46°$ and $BC = 4.6$ cm. Construct **two** different triangles from this information.

10  Construct a triangle $ABC$ with angle $A = 68°$ and angle $B = 49°$.

# 4.4 Quadrilaterals

A **quadrilateral** is a shape with four straight sides and four angles. You can investigate the sum of the angles of a quadrilateral in a way similar to the one you used for a triangle.

Draw a quadrilateral and label its angles $a$, $b$, $c$ and $d$, Figure 4.17. Tear off the corners and fit the angles together at a point, Figure 4.18. From Chapter 02, the angles at a point add up to 360°. This suggests that $a + b + c + d = 360°$ and that the angle sum of a quadrilateral is 360°.

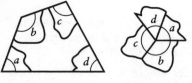 

| Figure 4.17 | Figure 4.18 | Figure 4.19 |

To *prove* this, draw a diagonal of a quadrilateral, dividing it into two triangles (see Figure 4.19). The angle sum of each triangle is 180°, so **the angle sum of the quadrilateral is 360°**.

A **square** is a quadrilateral with four equal sides and four right angles, Figure 4.20. A square has 4 lines of symmetry, Figure 4.21, and rotational symmetry of order 4.

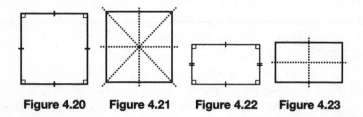

| Figure 4.20 | Figure 4.21 | Figure 4.22 | Figure 4.23 |

A **rectangle** has four right angles and its opposite sides are equal in length, Figure 4.22. A rectangle has 2 lines of symmetry, Figure 4.23, and rotational symmetry of order 2.

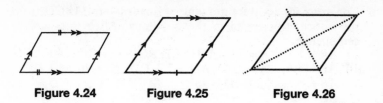

| Figure 4.24 | Figure 4.25 | Figure 4.26 |

A **parallelogram** has opposite sides which are both parallel and equal in length (see Figure 4.24). A parallelogram has no lines of symmetry and rotational symmetry of order 2.

A **rhombus**, Figure 4.25, is a parallelogram with equal sides. It has 2 lines of symmetry, Figure 4.26, and rotational symmetry of order 2.

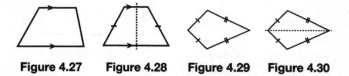

| Figure 4.27 | Figure 4.28 | Figure 4.29 | Figure 4.30 |

A **trapezium**, Figure 4.27, has one pair of parallel sides. It generally has no lines of symmetry and has rotational symmetry of order 1. An **isosceles trapezium**, Figure 4.28, in which the non-parallel sides are equal in length, has one line of symmetry.

A **kite**, Figure 4.29, has two pairs of adjacent sides with equal lengths, like two isosceles triangles with their bases joined. A kite has 1 line of symmetry, Figure 4.30, and rotational symmetry of order 1.

### EXERCISE 4.3

1 In the figure, work out the size of the angle marked $x$.

2 Write down the names of two quadrilaterals with rotational symmetry of order 2.

3 Write down the names of three quadrilaterals whose diagonals cross at right angles.

4 A quadrilateral has $m$ lines of symmetry. Write down all the possible values of $m$.

5 A quadrilateral has rotational symmetry of order $n$. Write down all the possible values of $n$.

6 Construct an accurate drawing of quadrilateral *ABCD* from the information given in the questions. All lengths are in centimetres.
  (a) *AB* = 3.7, ∠*A* = 78°, *AD* = 2.1, *BC* = 3.1, *CD* = 4.6.
  (b) *AB* = 4.9, ∠*A* = 114°, *BD* = 5.7, *CD* = 3.9, *BC* = 3.2.
  (c) *AB* = 4.1,   ∠*BAD* = 98°,   ∠*ABD* = 34°,   *CD* = 2.9, *BC* = 4.6.
  (d) *AB* = 3.8, *AD* = 2.9, *BD* = 4.5, ∠*BDC* = 31°, *CD* = 4.7.

## 4.5 Polygons

A **polygon** is a shape with three or more straight sides. To calculate the angle sum of a polygon, split it into triangles. For the five-sided polygon in Figure 4.31, choose any vertex, draw as many diagonals as you can from that vertex, in this case, 2, splitting the five-sided polygon into three triangles. The angle sum of a triangle is 180°, so the angle sum of the five-sided polygon is 3 × 180°, or 540°.

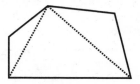

**Figure 4.31**

In general, the number of triangles into which you can split a polygon in this way is two less than the number of sides of the polygon.

The table shows the angle sums and the names of some polygons.

| Number of sides | Name | Number of triangles | Angle sum in degrees |
|---|---|---|---|
| 3 | Triangle | 1 | 180 |
| 4 | Quadrilateral | 2 | 360 |
| 5 | Pentagon | 3 | 540 |
| 6 | Hexagon | 4 | 720 |
| 8 | Octagon | 6 | 1080 |
| 10 | Decagon | 8 | 1440 |

A polygon whose sides and angles are equal is called **regular**. You can find the size of each angle of a regular polygon by dividing the sum of its angles by the number of its angles.

# 4.6 Interior and exterior angles

So far in your study of polygons, the angles you have been given or have had to find have been *inside* the polygon. These angles are called **interior angles**. If you extend a side, as in Figure 4.32, the angle *outside* the shape is called an **exterior angle**.

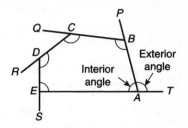

There is a surprising fact about the sum of the exterior angles of a polygon. Suppose that you are walking round the polygon, starting from *A* and facing *B*. Walk to *B*, and when you get there, turn from the

**Figure 4.32**

direction *BP* to the direction *BC*. Now walk to *C*, and when you get there, turn from the direction *CQ* to the direction *CD*. Continue in this way, until you come back to *A*, and you turn from *AT* to face *B*, and you are back where you started, and facing in the same direction. On the way, you have turned an angle equal to the angle sum of the exterior angles, but you have also turned through 360° because you are facing the same direction as before. Moreover, this argument applies to any polygon, not just the pentagon in Figure 4.32.

Therefore,

**the sum of the exterior angles of any polygon is 360°.**

This fact is often useful in questions involving regular polygons, because the exterior angles of a regular polygon are all equal.

---

**Example 4.6.1**

Work out the size of
(a) each exterior angle,
(b) each interior angle of a regular pentagon.

(a) A regular pentagon has five equal exterior angles whose sum is 360°, so each exterior angle is 360° ÷ 5 = 72°.
(b) At each vertex, interior angle + exterior angle = 180°.

Thus every interior angle is 180° − 72° = 108°.

**Example 4.6.2**

The size of each exterior angle of a regular polygon is 24°.
(a) Calculate the interior angle of the polygon.
(b) How many sides has the polygon?

(a) As each exterior angle is 24°, the interior angles are each $(180 - 24)°$, which is 156°.

(b) The sum of exterior angles is 360° and each exterior angle is 24°, so the number of sides is $360 \div 24 = 15$.

## EXERCISE 4.4

In Questions 1–3, work out the sizes of the angles marked with letters.

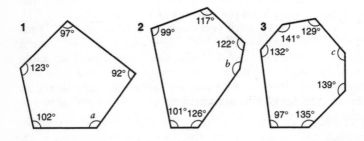

4   Work out the sum of the angles of a 12-sided polygon.

5   Work out the sum of the angles of a 15-sided polygon.

6   Work out the size of each angle of a regular pentagon.

7   Work out the size of each angle of a regular decagon.

8   Work out the size of each angle of a regular 18-sided polygon.

9   The size of an interior angle of a polygon is 109°. Work out the size of the exterior angle at the same vertex.

10   Work out the size of each exterior angle of a regular hexagon.

11   Work out the size of (a) each exterior angle and (b) each interior angle of a regular 12-sided polygon.

12   The size of each interior angle of a regular polygon is 140°. How many sides has the polygon?

13   The size of each interior angle of a regular polygon is 168°. How many sides has the polygon?

## 4.7 Symmetries of regular polygons

You already know that an equilateral triangle (a regular polygon with 3 sides) has 3 lines of symmetry and that a square (a regular polygon with 4 sides) has 4 lines of symmetry. Figure 4.33 shows the lines of symmetry drawn on regular polygons with 5, 6, 7 and 8 sides.

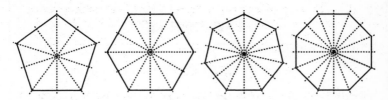

**Figure 4.33**

In general, the number of lines of symmetry of a regular polygon and the order of rotational symmetry of a regular polygon are both the same as the number of its sides.

## 4.8 Congruent shapes

If you make a tracing of one of the triangles in Figure 4.34 you will find that the tracing will fit exactly on top of each of the other triangles.

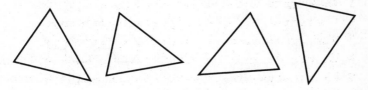

**Figure 4.34**

The triangles in Figure 4.34 are exactly the same shape and size and they are said to be **congruent**.

If you turn the tracing paper over and the tracing fits exactly on top of a triangle, that triangle is also congruent to the original one. Use tracing paper to check that the triangles in Figure 4.35 are also congruent.

Shapes other than triangles may also be congruent to each other.

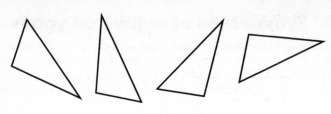

**Figure 4.35**

## EXERCISE 4.5

Write down the letters of the pairs of congruent shapes in the diagram.

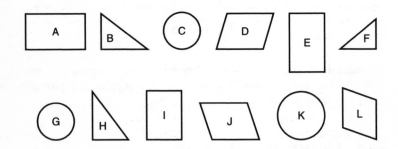

## 4.9 Tessellations

Figure 4.36 shows how regular hexagons can fit together without any gaps and without overlaps. Bees' honeycombs are an example of this.

Shapes which fit together like this are said to **tessellate** and the tiling patterns they make are called **tessellations**.

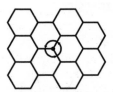

**Figure 4.36**

'Tessellate' and 'tessellations' are derived from the Latin word for the small square stones used to make mosaic patterns.

Each interior angle of a regular hexagon is 120° so three corners fit together to make 360°.

Figure 4.37 shows how equilateral triangles tessellate. (In fact, *all* triangles tessellate.) Each angle of an equilateral triangle is 60° so six corners fit together to make 360°.

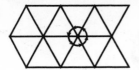

**Figure 4.37**

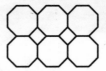

**Figure 4.38**

Regular octagons do not tessellate but a combination of regular octagons and squares will tessellate, as Figure 4.39 shows. Each interior angle of a regular octagon is 135° and two of these corners fit together with a right angle to make 360°. This does not guarantee that the regular octagons and squares will tessellate, but if it were not true, they certainly wouldn't tessellate.

There is a wide variety of beautiful tessellations, ranging from intricate Islamic mosaics to the work of the Dutch artist M. C. Escher.

## EXERCISE 4.6

1. Draw diagrams to show how each of these shapes tessellates. Show at least six shapes in each diagram.
   (a) isosceles triangle          (b) square
   (c) parallelogram               (d) kite

2. Without drawing, explain why rectangles tessellate.

3. Without drawing, explain why regular nonagons (nine-sided polygons) do not tessellate.

4. Draw a diagram with at least ten shapes to show a tessellation using a combination of regular hexagons and equilateral triangles.

5. Explain, without a drawing, why it may be possible to make a tessellation using a combination of equilateral triangles and regular 12-sided polygons. Then, using a drawing, show that the tessellation is possible.

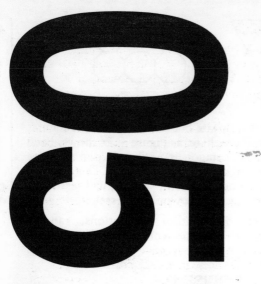

**05**

**decimals**

In this chapter you will learn:

- about place value in decimal numbers
- how to convert between decimals and fractions
- how to add and subtract decimals
- how to multiply and divide decimals
- about terminating and recurring decimals.

# 5.1 Introduction

The invention of decimal fractions, usually called simply decimals, was a gradual process. It started around the twelfth century and was completed in 1585, when Simon Stevin, a Flemish scientist, published a treatment called *La Disme* (English title: *The Art of Tenths*).

Stevin's symbolism was, however, very complicated and, although numbers written with a decimal point had first appeared in print in 1492, decimals were not generally adopted until the late seventeenth century. In fact, even today, the symbolism is not standardized. While the United Kingdom and the United States use a decimal point, many European countries use a comma instead.

Although Lord Randolph Churchill, Sir Winston's father, complained that he 'never could make out what those damned dots meant', the use of decimals brought many benefits, including making commercial processes easier and, in mathematics, the development of logarithms.

# 5.2 Place value

In Chapter 01, you saw how, in the decimal system, the place of a digit in a number tells you the value of that digit. The value of the digit in each of the first three columns is

**hundreds   tens   units**

Each value is $\frac{1}{10}$ of the value on its left. You can continue this pattern to the right of the units column with column values of **tenths, hundredths** and **thousandths**.

The value of each digit in the number 654.739, which is read as six hundred and fifty four **point** seven three nine, is shown in the table.

| hundreds | tens | units | | tenths | hundredths | thousandths |
|----------|------|-------|---|--------|------------|-------------|
| 6 | 5 | 4 | . | 7 | 3 | 9 |

The **decimal point** acts as a marker which separates the whole number part from the part which is less than 1.

For example, the value of the 7 is $\frac{7}{10}$ and the value of the 9 is $\frac{9}{1000}$.

The number 654.739 has three **decimal places**. The digit 7 is in the first decimal place; 3 is in the second decimal place and 9 is in the third.

With decimals which are less than 1, it is usual, but not essential, to write a zero on the left of the decimal point. Thus, you write 0.53 rather than .53 to draw attention to the decimal point.

If two decimal numbers have equal whole number parts, you can find which is larger by comparing their tenths digits, their hundredths digits and so on. For example, 4.23, which has a tenths digit of 2, is greater than 4.19, which has a tenths digit of 1. It does not matter that 4.19 has a greater hundredths digit than 4.23.

---

**Example 5.2.1**

Which decimal fraction is greater, 0.506 or 0.512?

Start from the decimal point and work to the right. The tenths digits are equal. 0.506 has a hundredths digit of 0 and 0.512 has a hundredths digit of 1. As 1 is greater than 0, 0.512 is greater than 0.506.

---

The symbol $>$ means 'is greater than'. Example 5.2.1 shows that $0.512 > 0.506$. The symbol $<$ means 'is less than'. Thus $0.506 < 0.512$.

## EXERCISE 5.1

1 Write these numbers in a table like that at the top of this page and, in each case, state the value of the digit 3.
   (a) 174.783   (b) 936.214   (c) 472.329   (d) 579.038

2 Write down the value of the underlined digit in each of these decimal numbers.
   (a) 2<u>3</u>.045   (b) 7.<u>8</u>1   (c) 0.8<u>92</u>   (d) 0.01<u>9</u>

3 (a) How many decimal places are there in the number 231.67?
   (b) Write down the digit which is in the second decimal place.

4 (a) How many decimal places are there in the number 0.609?
   (b) Write down the digit which is in the second decimal place.

5 Find the greater decimal fraction of each of these pairs.
   (a) 0.899, 0.901   (b) 0.702, 0.72
   (c) 0.01, 0.009   (d) 0.101, 0.909

6 Write down the largest number from this list.
   2.365     5.362     5.632     5.236     3.625

7 Write down the smallest number from this list.
   0.786      0.876      0.678      0.687      0.768

8 Write these numbers in order of size. Start with the smallest.
   1.32      1.302      1.203      2.31      1.23

9 Write these numbers in order of size. Start with the smallest.
   0.190      0.910      0.109      0.901      0.019

10 Insert either < or > between each of these pairs of number
   to give a true statement.
   (a) 4.76, 6.47      (b) 5.72, 5.27
   (c) 0.39, 0.4      (d) 0.03, 0.029

## 5.3 Converting decimals to fractions

You can use place value to convert decimals to fractions.

For example, $0.7 = \frac{7}{10}$ and $0.4 = \frac{4}{10}$.

The fraction $\frac{7}{10}$ is already in its simplest form, but you can simplify $\frac{4}{10}$ to get $0.4 = \frac{2}{5}$. (See Section 3.4.)

You can extend this method to numbers, such as 0.37, with two digits on the right of the decimal point.

$$0.37 = \frac{3}{10} + \frac{7}{100} = \frac{30}{100} + \frac{7}{100} = \frac{37}{100}$$

You can write decimals of this type straight down as 'hundredths'. You should always give your final answer in its simplest form. The fraction $\frac{37}{100}$ is already in its simplest form, but the simplest form can be a fraction with a denominator smaller than 100.

For example, $0.45 = \frac{45}{100}$, which is $\frac{9}{20}$ in its simplest form.

Similarly, numbers with three digits on the right of the decimal point may be written straight down as 'thousandths' and then simplified, if possible. For example, $0.139 = \frac{139}{1000}$, which cannot be simplified.

When you convert a decimal number greater than 1 to a mixed number, the whole number part stays the same and you convert the decimal fraction to a fraction. For example, $5.3 = 5\frac{3}{10}$.

## 5.4 Converting fractions to decimals 1

If the fraction has a denominator which is 10, 100, 1000, etc., you can use place value to convert the fraction directly to a decimal. For example, $\frac{27}{100} = 0.27$.

Even if the fraction does not have 10 or 100 or 1000 as its denominator, you can sometimes convert it to an equivalent fraction which does. (See Section 3.2.) For example, $\frac{3}{20} = \frac{15}{100} = 0.15$.

When you convert a mixed number to a decimal, the whole number part stays the same and you convert the fraction to a decimal fraction. For example, $4\frac{3}{50} = 4\frac{4}{100} = 4.06$.

The methods you have met so far in this section apply only to fractions with a denominator of 10 or 100 or 1000 or with a denominator which is a factor of one of these numbers. Later in this chapter you will meet a method which may be used to convert any fraction to a decimal.

### EXERCISE 5.2

1 Convert each of these decimals to a fraction in its simplest form.
  (a) 0.6     (b) 0.1     (c) 0.48     (d) 0.39
  (e) 0.09    (f) 0.179   (g) 0.125   (h) 0.003

2 Convert each of the following to a mixed number with the fraction in its simplest form.
  (a) 3.8     (b) 4.22    (c) 54.75    (d) 23.048

3 Convert each of these fractions to a decimal.

  (a) $\frac{3}{10}$     (b) $\frac{71}{100}$     (c) $\frac{7}{100}$     (d) $\frac{387}{1000}$

  (e) $\frac{17}{1000}$    (f) $\frac{9}{1000}$    (g) $\frac{2}{5}$     (h) $\frac{19}{20}$

  (i) $\frac{17}{25}$     (j) $\frac{11}{50}$    (k) $\frac{237}{500}$    (l) $\frac{7}{200}$

4 Convert each of these mixed numbers to a decimal.

  (a) $6\frac{1}{10}$     (b) $9\frac{3}{100}$    (c) $5\frac{41}{1000}$    (d) $1\frac{9}{20}$

  (e) $8\frac{2}{25}$    (f) $12\frac{37}{200}$

## 5.5 Addition and subtraction

If you line up the decimal points underneath each other, adding and subtracting decimals is very similar to adding and subtracting whole numbers.

For example,

with whole numbers
$$\begin{array}{r} 354 \\ 2432\ + \\ \hline 2786 \end{array}$$

and with decimals
$$\begin{array}{r} 3.54 \\ 24.32\ + \\ \hline 27.86 \end{array}$$

**Example 5.5.1**

Work out   (a)  $9.7 + 6.83$,   (b)  $12.4 + 0.72$,
(c)  $0.71 + 6.092$.

| (a)  9.7 | (b)  12.4 | (c)  0.71 |
|---|---|---|
| 6.83 + | 0.72 + | 6.092 + |
| 16.53 | 13.12 | 6.802 |

If you find it helpful, you can put zeros in to the right of the last digit, so that all the numbers have the same number of decimal places. For example, in part (a), you can write 9.7 as 9.70, which may help you to align the decimals.

**Example 5.5.2**

Work out   (a)  $6.9 - 2.4$,   (b)  $12.3 - 4.7$,
(c)  $19.04 - 6.2$.

| (a)  6.9 | (b)  12.3 | (c)  19.04 |
|---|---|---|
| 2.4 − | 4.7 − | 6.2 − |
| 4.5 | 7.6 | 12.84 |

If the calculations are more difficult than those given, you should use your calculator.

### EXERCISE 5.3

1  Work out the answers to the following questions.
   - (a)  $3.2 + 4.7$
   - (b)  $8.6 + 4.52$
   - (c)  $9.82 + 0.48$
   - (d)  $23.8 + 0.27$
   - (e)  $9.7 - 3.4$
   - (f)  $24.3 - 7.5$
   - (g)  $26.37 - 19.2$
   - (h)  $32.4 - 17.57$
   - (i)  $6 + 4.7$
   - (j)  $9 - 6.8$
   - (k)  $12 - 0.12$
   - (l)  $7.9 - 4.53$
   - (m)  $7.6 - 3.749$
   - (n)  $3.86 - 0.372$
   - (o)  $8.4 + 3.685$

2  Work out the difference between 8.43 and 9.2.

3  The judges awarded a skater the scores 5.8, 5.7, 6, 5.4, 6 and 5.9. Work out the skater's total score.

4  The runner-up in a 100-metres race had a time of 10.37 seconds, which was 0.09 seconds behind the winner. Find the winner's time.

# 5.6 Multiplication of decimals

To multiply a decimal by a whole number less than 10, you should set out your working so that the decimal point in the answer is directly below the decimal point in the question. Then you can do the multiplication as if you were multiplying two whole numbers together. For example, the working for $3.6 \times 4$ is shown on the right.

$$
\begin{array}{r}
3.6 \\
4 \times \\
\hline
14.4 \\
\hline
\end{array}
$$

To multiply decimals by 10 or 100 or 1000, there are some simple rules, which you will remember more easily if you see how they are obtained. The diagram illustrates $32.69 \times 10$.

| Hundreds | Tens | Units | • | Tenths | Hundredths |
|----------|------|-------|---|--------|------------|
|          | 3    | 2     | • | 6      | 9          |
| 3        | 2    | 6     | • | 9      |            |

The digits have moved one place to the left, as their place values are 10 times bigger but, when the result is written down without the column headings as $32.69 \times 10 = 326.9$, it looks as if the decimal point has moved one place to the right.

Using one of these rules, you can write down the result of multiplying a number by 10 without working. For example, $5.732 \times 10 = 57.32$. Your answer may be a whole number. For example, $38.7 \times 10 = 387$.

Multiplication of a decimal by 100 may be approached in a similar way. For example, $4.173 \times 100 = 417.3$.

The multiplication of two decimal numbers may be introduced by converting them both to fractions. For example, to work out $0.3 \times 0.7$, Write it as $\frac{3}{10} \times \frac{7}{10}$. Using the method you met in Section 3.7, the result is $\frac{21}{100}$, which, as a decimal is 0.21, that is, $0.3 \times 0.7 = 0.21$.

Notice that $3 \times 7 = 21$. Notice also that 0.3 and 0.7 each have one decimal place and that 0.21 has two decimal places, that is, the number of decimal places in the answer is the same as the total number of decimal places in the question. This approach enables you to write down the answers to similar products, such as $0.6 \times 0.5 = 0.30$, which may be written as 0.3, and $0.3 \times 0.2 = 0.06$ (not 0.6).

---

**Example 5.6.1**

Write down the answer to  (a) $0.6 \times 0.9$,  (b) $0.5 \times 0.8$, (c) $0.3 \times 0.3$.

(a) $0.6 \times 0.9 = 0.54$.  (b) $0.5 \times 0.8 = 0.40$ (or 0.4).
(c) $0.3 \times 0.3 = 0.09$.

---

To work out a more difficult product, such as $0.4 \times 0.27$, you could convert each decimal to a fraction and multiply them: $\frac{4}{10} \times \frac{27}{100} = \frac{108}{1000}$, which is 0.108 as a decimal, so $0.4 \times 0.27 = 0.108$.

There is a better method: $4 \times 27 = 108$. Also, 0.4 has one decimal place, 0.27 has two decimal places and 0.108 has three decimal places. So, again, the number of decimal places in the answer is the same as the total number of decimal places in the question.

This suggests a general method for multiplication involving decimals.

*Step 1*  Ignore the decimals and multiply the numbers together.
*Step 2*  Find the total number of decimal places in the numbers you are multiplying.
*Step 3*  The answer must have the same number of decimal places.

Here the steps are used to calculate $0.07 \times 0.008$.

*Step 1*  The calculation is $7 \times 8 = 56$.
*Step 2*  0.07 has two decimal places and 0.008 has 3 decimal places, so the answer must have $2 + 3 = 5$ decimal places.
*Step 3*  The answer has five decimal places, therefore $0.07 \times 0.008 = 0.00056$.

---

**Example 5.6.2**

Write the answers to  (a) $0.02 \times 0.04$,  (b) $0.05 \times 0.8$.

(a) $0.02 \times 0.04 = 0.0008$.  (b) $0.05 \times 0.8 = 0.040$ or 0.04.

---

If one of the numbers you want to multiply is a whole number, you must remember that a whole number has **no** decimal places.

The result in Example 5.6.1, $24.71 \times 3 = 74.13$, is consistent with this.

If the calculations are any more complicated than those given so far, you are advised to use a calculator.

### EXERCISE 5.4

1 Work out
   (a) $4.7 \times 6$ (b) $17.8 \times 5$ (c) $18.37 \times 8$ (d) $0.403 \times 3$

2 Write down the answers to
   (a) $8.21 \times 10$      (b) $0.6 \times 10$      (c) $0.47 \times 10$
   (d) $0.702 \times 10$      (e) $28.3 \times 10$      (f) $0.0041 \times 10$
   (g) $5.218 \times 100$      (h) $0.83 \times 100$      (i) $0.4 \times 100$
   (j) $0.0007 \times 100$      (k) $46.36 \times 1000$      (l) $3.9 \times 1000$

3 Copy and complete
   (a) $0.21 \times \ldots = 21$      (b) $6.71 \times \ldots = 67.1$
   (c) $7.98 \times \ldots = 7980$

4 Write down the answers to
   (a) $0.9 \times 0.7$ (b) $0.6 \times 0.5$ (c) $0.2 \times 0.4$
   (d) $0.1 \times 0.3$ (e) $0.07 \times 0.3$ (f) $0.03 \times 0.02$

5 Work out
   (a) $3.7 \times 1.4$      (b) $0.28 \times 0.41$      (c) $6.8 \times 0.37$
   (d) $0.084 \times 0.47$      (e) $9.3 \times 37$      (f) $0.049 \times 23$

6 1 litre $= 1.76$ pints. How many pints are there in 6 litres?

7 1 gallon $= 4.5$ litres. How many litres are there in 8 gallons?

8 1 metre $= 100$ centimetres. How many centimetres are there in 8.32 metres?

## 5.7 Division of decimals

To divide a decimal by a whole number less than 10, you should, as with multiplication, set out your working so that the decimal point in the answer is directly below the decimal point in the question.

Then you can do the division as if you were dividing one whole number by another. For example, the working for $13.5 \div 5$ is shown on the right.

$$5) \overline{\underset{2.7}{13.5}}$$

To divide decimals by 10 or 100 or 1000 you can use simple rules which are the opposite of the ones you used for multiplication by these numbers. The diagram below illustrates $574.6 \div 10$.

| Hundreds | Tens | Units | • | Tenths | Hundredths |
|----------|------|-------|---|--------|------------|
| 5 | 7 | 4 | • | 6 | |
| | 5 | 7 | • | 4 | 6 |

The *digits* have moved one place to the *right*, as their place values are one tenth of what they were but, when the result is written down without the column headings as $574.6 \div 10 = 57.46$, it looks as if the *decimal point* has moved one place to the *left*.

Using one of these rules, you should be able to write down the result of dividing a decimal by 10 without any working. For example, $34.19 \div 10 = 3.419$. You may need to write zeros to show that columns are empty. For example, $0.93 \div 10 = 0.093$.

To divide a decimal by 100 you move the decimal point *two* places to the *left* and to divide a decimal by 1000, you move the decimal point *three* places to the *left*.

To divide by a decimal, write the division as a fraction, as the line in a fraction may be thought of as a division sign. (See Section 3.5.) Then, by multiplying by 10 or 100 or 1000, convert it to an equivalent fraction with a whole number for the denominator, as you already know how to divide by a whole number. The numerator does not have to be a whole number.

For example, to divide 0.12 by 0.3, write it as a fraction $\frac{0.12}{0.3}$. Then multiply both the numerator and denominator by 10 and obtain $\frac{1.2}{3}$, which you can evaluate to get the answer 0.4.

If the calculations are any more complicated than those given so far, you are advised to use a calculator.

---

**Example 5.7.1**

Work out $2.8 \div 0.04$.

As a fraction, $2.8 \div 0.04 = \frac{2.8}{0.04}$. Multiplying both the numerator and denominator by 100, $\frac{2.8}{0.04} = \frac{280}{4}$. Finally, $\frac{280}{4} = 70$.

# EXERCISE 5.5

1  Work out
   (a) $9.2 \div 4$   (b) $19.23 \div 3$   (c) $2.382 \div 6$   (d) $0.469 \div 7$

2  Write down the answers to
   (a) $37.5 \div 10$      (b) $457.2 \div 10$      (c) $0.093 \div 10$
   (d) $7.13 \div 10$      (e) $0.34 \div 10$       (f) $652.8 \div 100$
   (g) $2.5 \div 100$      (h) $0.37 \div 100$      (i) $0.003 \div 100$
   (j) $17.2 \div 100$     (k) $33.8 \div 1000$     (l) $0.071 \div 1000$

3  Copy and complete
   (a) $3.7 \div \ldots = 0.37$      (b) $0.9 \div \ldots = 0.009$
   (c) $7.9 \div \ldots = 0.079$

4  Work out
   (a) $39.92 \div 2$      (b) $199.5 \div 7$       (c) $0.624 \div 4$
   (d) $7.28 \div 8$       (e) $257.6 \div 70$      (f) $10.8 \div 300$
   (g) $43.75 \div 50$     (h) $2.618 \div 400$

5  Work out
   (a) $0.36 \div 0.4$      (b) $3 \div 0.2$         (c) $4.8 \div 0.06$
   (d) $0.021 \div 0.7$     (e) $9 \div 0.03$        (f) $4 \div 0.08$
   (g) $7.38 \div 0.3$      (h) $3.384 \div 0.09$

6  A length of cable 18.48 metres long is cut into 4 equal pieces.
   Work out the length of each piece.

7  A pile of 100 sheets of paper is 2.1 centimetres high. Work out
   the thickness of each sheet of paper.

8  In completing one lap, a runner covers 0.4 kilometre. How
   many laps does she complete in a 10-kilometre race?

# 5.8 Converting fractions to decimals 2

In Section 5.4, you saw how to convert certain types of fractions
to decimals. There is a general method which you can use to
convert *any* fraction to a decimal. You have to think of the line
in a fraction as a division sign. So, for example, $\frac{3}{4}$ means $3 \div 4$.

You can write as many zeros
as you like on the right of the
decimal point.

$$\begin{array}{r} 4)\overline{3.00} \\ \hline 0.75 \end{array}$$

The decimal 0.75 is called a **terminating decimal**, because it
divides out exactly and the decimal terminates or stops, after
two decimal places in this case.

To convert a mixed number to a decimal, you leave the whole number unchanged and use the division method to convert the fraction.

---

**Example 5.8.1**

Convert $3\frac{7}{8}$ to a decimal.

If you divide 7 by 8 you get 0.875, so $\frac{7}{8} = 0.875$.

Therefore $3\frac{7}{8} = 3 + 0.875 = 3.875$.

---

Not all fractions give terminating decimals when you convert them. Using the division method with $\frac{2}{3}$, for example, you find that the decimal repeats itself, and $\frac{2}{3} = 0.666\ldots$, where the three dots means that the sixes repeat for ever.

This kind of decimal, which keeps on repeating, is called a **recurring decimal**. To avoid writing down the same figure repeatedly, you can write it in a shortened form, using dot notation, with a dot above the repeating figure. The decimal $0.666\ldots$ then becomes $0.\dot{6}$.

---

**Example 5.8.2**

Convert $\frac{5}{6}$ to a recurring decimal and write it in dot notation.

$$6)\,\overline{5.0000}$$
$$0.8333$$

$$\frac{5}{6} = 0.8\dot{3}.$$

---

Sometimes, more than one number repeats. When this happens, dots are placed above the first and last figures in the repeating group. The fraction $\frac{3}{11}$, for example, written as a decimal is $0.272\,727\ldots$, which you write as $0.\dot{2}\dot{7}$. Similarly, $\frac{17}{37} = 0.459\,459\,459\ldots = 0.\dot{4}5\dot{9}$.

When writing recurring decimals in dot notation, you must make sure that you place the dots over the first and last of the group of numbers that repeat. So, for example, $0.58216216216\ldots$ is written as $0.58\dot{2}1\dot{6}$.

When you convert a mixed number to a recurring decimal, you place the dots above the decimal part and leave the whole number unchanged, so that, in dot notation, 7.7777... is 7.7̇.

In Chapter 03, Section 3.3, you saw how to compare the sizes of fractions using equivalent fractions. Although this method will work for any fractions, it would not be ideal for comparing fractions with large denominators such as $\frac{15}{19}$ and $\frac{18}{23}$. It is much easier to convert each of these to a decimal using a calculator. The full calculator reading for $15 \div 19$ is 0.789 473 684, and for $18 \div 23$ it is 0.782 608 696, but you only need to go as far as the third decimal place to see that the first decimal is greater than the second, so $\frac{15}{19} > \frac{18}{23}$.

## EXERCISE 5.6

**1** Convert each of these fractions to decimals and state whether it is a terminating decimal or a recurring decimal.

(a) $\frac{1}{4}$    (b) $\frac{1}{3}$    (c) $\frac{3}{8}$

(d) $\frac{4}{9}$    (e) $\frac{7}{8}$

**2** Using a calculator, convert each of these fractions to a recurring decimal and write it in dot notation.

(a) $\frac{9}{11}$    (b) $\frac{1}{11}$    (c) $\frac{5}{12}$    (d) $\frac{1}{12}$

(c) $\frac{19}{33}$    (f) $\frac{3}{37}$    (g) $\frac{23}{66}$    (h) $\frac{39}{74}$

**3** Write each of these recurring decimals in dot notation.

(a) 0.8888...      (b) 0.712 121 2... (c) 0.367 367...
(d) 0.456 896 89... (e) 18.018 18...    (f) 0.564 356 43...
(g) 9.3999...        (h) 263.6363...

**4** Insert < or > between each of these pairs of fractions to make a true statement. You should use a calculator.

(a) $\frac{7}{11}, \frac{8}{13}$    (b) $\frac{14}{19}, \frac{17}{23}$    (c) $\frac{6}{13}, \frac{7}{15}$    (d) $\frac{13}{17}, \frac{16}{21}$

**5** Use a calculator to put these numbers in order, smallest first.

(a) $5.8, 5\frac{7}{9}, 5.77, 5.78, 5\frac{17}{22}$

(b) $3.1, 3\frac{1}{11}, 3.01, 3.09, 3\frac{2}{21}$

(c) $4\frac{1}{12}, 4.084, 4.083, 4\frac{2}{23}, 4.085$

**6** Find a decimal with three decimal places which is between $\frac{16}{31}$ and $\frac{17}{33}$. Convert your decimal to a fraction in its simplest form.

7 (a) Convert these fractions to recurring decimals.
$\frac{1}{7}, \frac{2}{7}, \frac{3}{7}, \frac{4}{7}, \frac{5}{7}, \frac{6}{7}$

(b) How many figures are there in each of the recurring groups?

(c) Comment on the figures in the recurring groups.

8 (a) Which denominators of the fractions $\frac{1}{2}, \frac{1}{3}, \frac{1}{4}, \ldots, \frac{1}{24}, \frac{1}{25}$ when converted to decimals, give rise to terminating decimals?

(b) Comment on the numbers in your answer to (a).

(Hint: find the factors of each denominator.)

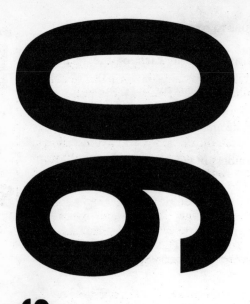

# 06

## statistics 1

**In this chapter you will learn:**

- about the collection of statistical data
- how to draw and use pictograms, bar charts, pie charts and line graphs
- how to draw and use scatter graphs
- about correlation
- about discrete and continuous data.

# 6.1 Introduction

Statistics is the branch of mathematics concerned with the collection, organization and analysis of data, often, but not always, numerical data. This chapter deals with the first two of these three aspects of statistics, with particular emphasis on the presentation of data. The analysis of data, using a variety of statistical averages, will be found in Chapter 22.

Starting with pictorial records on the walls of cave dwellings, the collection and recording of data is as old as civilization itself. Before 3000 BC, the Babylonians gathered information about crops and trade and so, centuries later, did the Egyptians and the Chinese. The Greeks held censuses, for taxation purposes, as early as 594 BC but the Romans were the first to gather a wide range of population, financial and agricultural data about their empire. The word 'statistics' itself is derived from the Latin word *statisticus,* meaning 'of state affairs'.

The first real statistical study of population, *Observations on the London Bills of Mortality,* was published in 1662 and statistical methods subsequently assumed increasing importance, especially in the natural and social sciences.

Today, important decisions in many fields are made on the basis of statistics, although this reliance does have its critics. Benjamin Disraeli believed 'There are three kinds of lies: lies, damned lies and statistics.' Rex Todhunter Stout held a similar view: 'There are two kinds of statistics, the kind you look up and the kind you make up.'

# 6.2 Collection of data

You can collect statistical data in three ways: by referring to existing source material, such as newspapers; by carrying out an experiment; and by conducting a **survey**. A survey might involve a whole population, as in a national census, but, to save time and money, it is more likely that you would survey part of the whole population, called a **sample**.

A data collection sheet or tally chart is used to organize the data from a survey. You can use it to collate data from completed questionnaires or you can record information directly onto it.

For example, you can record the colours of 30 cars in a table.

| Colour | Tally | Frequency |
|--------|-------|-----------|
| Red | ⅢⅢ ⅢⅢ Ⅲ | 13 |
| Blue | ⅢⅡ | 4 |
| White | ⅢⅢ | 5 |
| Black | ⅢⅢ Ⅲ | 8 |

To make the tally marks easier to add up, group them in fives, drawing the fifth mark across the first four. The total of the tally marks for each colour is called the **frequency**, which is shown in the right-hand column. With this extra column, the table is called a **frequency table**. A frequency table need not have a column for the tally marks.

## 6.3  Pictograms

Data are often easier to understand if you present them in diagrammatic form. One example is a **pictogram** (or pictograph), in which you use pictures to show information. The frequency table below shows the numbers of cars using a car park on five days. You can illustrate the data in a pictogram by using the symbol 🚗 to represent 10 cars.

You can find how many symbols you need to represent the number of cars on each day by dividing the number of cars by 10 to get 3, 5, 4.5, 4.7 and 2.2. The pictogram is shown in Figure 6.1.

Only complete symbols are needed to represent the numbers of cars on Monday and Tuesday but, for each of the other days, parts of a symbol are needed in addition to complete symbols. You must use your judgement to draw the parts of symbols as accurately as possible.

| Day | Number of cars | Cars using a car park | | | | |
|-----|----------------|-----------------------|---|---|---|---|
| Monday | 30 | 🚗 🚗 🚗 | | | | |
| Tuesday | 50 | 🚗 🚗 🚗 🚗 🚗 | | | | |
| Wednesday | 45 | 🚗 🚗 🚗 🚗 🚗 | | | | |
| Thursday | 47 | 🚗 🚗 🚗 🚗 🚗 | | | | |
| Friday | 22 | 🚗 🚗 ◄ | | | | |

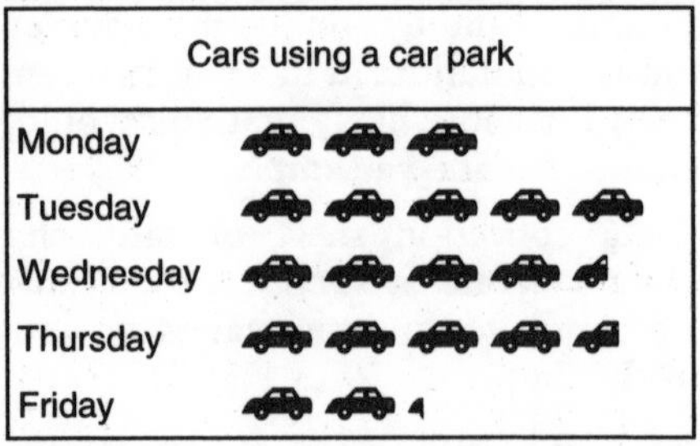

**Figure 6.1**

## 6.4 Bar charts

Pictograms are attractive but time-consuming to draw and it can be awkward to draw parts of a symbol. A **bar chart** is a simpler way of showing information. Figure 6.2 is a bar chart showing the information about the parked cars. In this case the bars are drawn vertically.

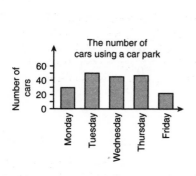

**Figure 6.2**                    **Figure 6.3**

You could, if you wish, draw the bars horizontally. In that case, the bar chart showing the same information is shown in Figure 6.3.

Which version you choose may ultimately depend on which you prefer, and the space which you have available.

### EXERCISE 6.1

1 Thirty two men were asked how they travelled to work. Their results are shown in the table. Complete the table.

| How they travel | Tally | Frequency |
|---|---|---|
| Walk | JHT JHT IIII | |
| Car | IIII | |
| Bus | JHT | |
| Train | JHT IIII | |

2 The numbers of children in 26 families are given below.

3 5 1 1 2 4 1 2 2 3 1 4 1
3 2 5 2 1 2 3 1 4 3 1 2 4

Collect this information in a frequency table.

3  The dress sizes of 25 women are shown in the frequency table.

| Dress size | 10 | 12 | 14 | 16 | 18 | 20 |
|---|---|---|---|---|---|---|
| Number of women | 1 | 5 | 7 | 6 | 4 | 2 |

Draw a pictogram to show this information. Use a symbol to represent one woman.

4  The numbers of students in three classes are shown in the table.

| Class | Number of students |
|---|---|
| History | 20 |
| Geography | 23 |
| Art | 31 |

Draw a pictogram to show this information. Use one symbol to represent 5 students.

5  The numbers of televisions in 100 houses are shown in the table.

| Number of televisions | Number of houses |
|---|---|
| 0 | 2 |
| 1 | 35 |
| 2 | 39 |
| 3 | 24 |

Draw a pictogram to show this information. Use a symbol to represent 10 houses.

6  The pictogram shows numbers of aircraft in four airlines.

| Airline | Number of aircraft |
|---|---|
| Fresh Air | ✈ ✈ |
| Air Britain | ✈ ✈ ، |
| Easy Flight | ✈ ✈ ✈ ✈ ✈ |
| Superjet | ✈ ✈ ✈ ✈ ✈ ✈ |

✈ represents 10 aircraft. Find the number of aircraft in each fleet.

7 The table shows the numbers of goals scored by ice hockey teams.

| Number of goals | 0 | 1 | 2 | 3 | 4 | 5 |
|---|---|---|---|---|---|---|
| Number of teams | 3 | 6 | 5 | 3 | 2 | 1 |

Draw a bar chart to show this information.

8 The passenger capacities of five types of Boeing aircraft are shown in the table. Draw a bar chart to illustrate this information.

| Type of aircraft | 737 | 747 | 757 | 767 | 777 |
|---|---|---|---|---|---|
| Passenger capacity | 106 | 426 | 195 | 252 | 281 |

9 100 men and 100 women in each of six age groups were asked if they wore spectacles. The numbers of men and women who said yes are shown in the table. Draw a bar chart to show this information.

| Age group (years) | 15–24 | 25–34 | 35–44 | 45–54 | 55–64 | 65+ |
|---|---|---|---|---|---|---|
| Number of men | 31 | 33 | 38 | 74 | 88 | 90 |
| Number of women | 37 | 42 | 44 | 79 | 87 | 89 |

10 The bar chart shows the lengths, in years, of the reigns of six British monarchs. Find the length of each of their reigns.

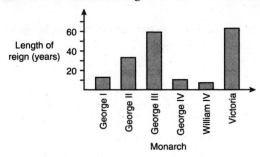

## 6.5 Pie charts

Information is often illustrated by means of **pie charts.** Pie charts show the proportions in which a quantity is shared out. You can draw a pie chart for the data about the car colours on page 64. The 30 cars are shown by a full circle, with an angle of 360° at its centre. One car is shown by an angle of $360° \div 30 = 12°$. The 13 red cars are shown by an angle of $13 \times 12°$, that is, 156°. You find the other angles similarly, getting 48°, 60° and 96° for the blue, white and black cars. With a protractor, draw the pie chart in Figure 6.4 with these angles at its centre. The 'slices' into which the circle is divided are called **sectors**.

The pie chart on its own does not tell you how many cars there were of each colour; it shows the proportion with each colour. An alternative method for finding the angles is to use fractions. For example, 13 of the 30 cars are red, so the angle representing these cars is $\frac{13}{30}$ of 360°, that is $\frac{13}{30} \times 360°$, which is 156°.

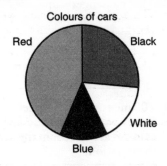

Colours of cars

**Figure 6.4**

The first method is more straightforward to use if the total number of items is a factor of 360. If the numbers are more awkward the fraction method is easier. Where appropriate, give angles to the nearest degree.

---

**Example 6.5.1**

The pie chart shows information about eye colours of 150 people.

(a) What fraction of the people have grey eyes? Give your fraction in its simplest form.

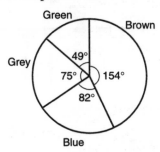

(b) How many people have brown eyes?

(a) The fraction is $\frac{75}{360}$ which is $\frac{5}{24}$ in its simplest form.

(b) Number of people with brown eyes $= \frac{154}{360} \times 150 = 64$.
This answer is given to the nearest whole number.

## EXERCISE 6.2

1 The table shows how a woman spent 24 hours. Draw a pie chart to show this information.

| Activity | Sleep | Work | Leisure | Meals | Travel |
|---|---|---|---|---|---|
| Number of hours | 8 | 7 | 6 | 1 | 2 |

2 The favourite types of food of 60 people are shown in the table. Draw a pie chart to show this information.

| Type of food | English | Italian | Indian | Chinese | Greek |
|---|---|---|---|---|---|
| Number of people | 11 | 9 | 19 | 14 | 7 |

3 A football team played 38 matches. They won 20, drew 11 and lost 7. Draw a pie chart to show this information.

4 Helen watched television for 3 hours 15 minutes. The figure gives information about the programmes she watched.

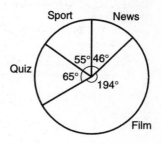

(a) How long did the news last?
(b) How long did the film last?
(c) For what fraction of her time did Helen watch sport? Give your answer in its simplest form.

# 6.6 Line graphs

The points plotted on the grid in Figure 6.5 show the temperature every 4 hours throughout a day.

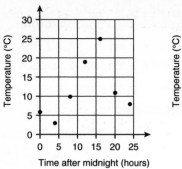

**Figure 6.5**

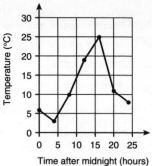

**Figure 6.6**

As the temperature changes gradually, you can join the points with straight lines. You then obtain the **line graph** in Figure 6.6.

Line graphs are often used to show data obtained over a period. You can use a line graph to read off values between those actually plotted. For example, from Figure 6.6, you can estimate the temperature at 2 p.m. as 22°C, but you must recognize that this can only be an estimate.

Joining points with straight lines is only appropriate when a quantity is changing gradually. You have to be even more careful about predicting future data by extending such a line graph.

The points plotted on the grid in Figure 6.7 show the average number of hours of sunshine each day in Budapest in six months of the year.

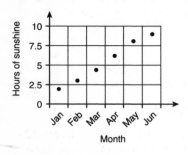

**Figure 6.7**

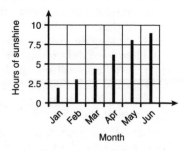

**Figure 6.8**

Points in between the plotted points have no meaning. A bar chart or a **vertical line graph,** as in Figure 6.8, would be a suitable way of showing this information.

When a quantity is changing gradually, rather than fluctuating, the plotted points may be joined either with straight lines or with a curve.

### Example 6.6.1

The temperature, in °C, at five minute intervals of a pan of hot water cooling down is shown in the table.

| Time (min) | 0 | 5 | 10 | 15 | 20 | 25 | 30 |
|---|---|---|---|---|---|---|---|
| Temperature (°C) | 100 | 60 | 42 | 33 | 28 | 25 | 23 |

(a) Plot the values in the table and join the points with a curve.
(b) Use your graph to estimate
  (i) the temperature of the water after 9 minutes,
  (ii) the time when the temperature was 66°C.

(a) The graph is an example of a cooling

(b) (i) Reading up from 9 on the time axis and across to the temperature axis, the temperature at 9 minutes was 44°C.

  (ii) Reading across from 66 on the temperature axis and down to the time axis, the temperature was 66°C after 4 minutes.

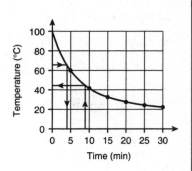

### EXERCISE 6.3

1 The height, in centimetres, measured every two days over a 12-day period, of a plant grown from seed is shown in the table.

| Time (days) | 0 | 2 | 4 | 6 | 8 | 10 | 12 |
|---|---|---|---|---|---|---|---|
| Height (cm) | 0 | 2 | 5 | 8 | 11 | 13 | 14 |

(a) Draw a line graph to show this information.
(b) Use your graph to estimate the height of the plant after five days.

2 A tray of water was placed in the freezing compartment of a fridge. In the next hour, the temperature, in °C, of the water was measured every ten minutes. The temperatures are shown in the table.

| Time (min) | 0 | 10 | 20 | 30 | 40 | 50 | 60 |
|---|---|---|---|---|---|---|---|
| Temperature (°C) | 21 | 13 | 7 | 5 | 4 | 3 | 2 |

(a) Draw a cooling curve for the water.
(b) Use your graph to estimate
   (i) the temperature after 5 minutes,
   (ii) the time when the temperature was 9°C.

3 A girl's height, in centimetres, measured every two years up to the age of fourteen is shown in the table.

| Age (years) | 0 | 2 | 4 | 6 | 8 | 10 | 12 | 14 |
|---|---|---|---|---|---|---|---|---|
| Height (cm) | 54 | 67 | 89 | 105 | 115 | 120 | 124 | 137 |

(a) Draw a line graph to show this information.
(b) Use your graph to estimate her height when she was aged five.

4 The population, to the nearest million, of the USA every 20 years from 1900 to 1980 is shown in the table.

| Year | 1900 | 1920 | 1940 | 1960 | 1980 |
|---|---|---|---|---|---|
| Population (millions) | 76 | 106 | 132 | 179 | 227 |

(a) Draw a line graph to show this information.
(b) Use your graph to estimate the population in 1930, and the year in which the population reached 200 million.

5 London's average monthly rainfall, in millimetres, in the first six months of the year is shown in the table.

| Month | Jan | Feb | Mar | Apr | May | Jun |
|---|---|---|---|---|---|---|
| Average monthly rainfall (mm) | 45 | 28 | 40 | 39 | 50 | 48 |

Draw a vertical line graph to show this information.

# 6.7 Scatter graphs

Scatter graphs show if there might be a relationship between two sets of data. For example, to investigate the relationship between women's heights and their weights, the data shown in the table were collected.

| Height (cm) | 167 | 152 | 175 | 163 | 164 | 169 | 155 | 171 | 160 | 156 |
|---|---|---|---|---|---|---|---|---|---|---|
| Weight (kg) | 66 | 54 | 72 | 63 | 67 | 69 | 57 | 72 | 64 | 61 |

You can record the data in the table on a grid, using height on one axis and weight on the other. This kind of graph, shown in Figure 6.9, is called a **scatter graph**.

Notice the zig-zags on the axes. They are put there to show that the axes are not graduated evenly up from zero.

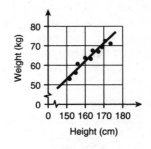

**Figure 6.9**

You can, by eye, draw a straight line, called a line of best fit, which passes near all these points. This line does not have to go through any of the points, although it can. There should be roughly equal numbers of points on each side of the line.

If you can draw a line of best fit, then there may be a relationship, called a **correlation,** between women's heights and their weights.

Weight increases as height increases; when one quantity increases as the other increases, the correlation is called **positive correlation.**

Correlation in which one quantity decreases as the other increases is called **negative correlation.** Figure 6.10 shows negative correlation.

Scatter graphs can show not only whether there is any correlation but also, if there is, how close the correlation is. On the scatter graph Figure 6.10, for example, all the points are close to the line of best fit and so this could be described as **high,** or strong, positive correlation.

When the scatter diagram has no pattern to it, as in Figure 6.11, there is no correlation between the two sets of data.

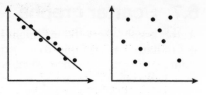

**Figure 6.10**          **Figure 6.11**

If all the points lie on a line, the correlation is called **perfect**. If the points are not close to the line of best fit, the correlation is called **low**, or weak. Figure 6.12 and Figure 6.13 show low positive correlation and low negative correlation.

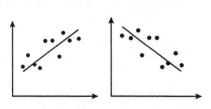

**Figure 6.12**          **Figure 6.13**

---

**Example 6.7.1**

The table shows the engine size in cubic centimetres (cc) and the fuel consumption in miles per gallon (mpg) of ten cars.

| Engine size (cc) | 1000 | 1100 | 1300 | 1400 | 1600 |
|---|---|---|---|---|---|
| Fuel consumption (mpg) | 61 | 59 | 54 | 52 | 51 |

| Engine size (cc) | 1800 | 2000 | 2500 | 2700 | 2900 |
|---|---|---|---|---|---|
| Fuel consumption (mpg) | 46 | 43 | 40 | 37 | 35 |

(a) Draw a scatter graph to show this information.
(b) Draw a line of best fit on your scatter graph.
(c) Describe the correlation.
(d) Estimate the fuel consumption of a car with a 2300 cc engine.

(a) and (b) These are in the diagram.
(c) There is a high negative correlation.
(d) Using the line of best fit, and following the arrow shown in the diagram, the petrol consumption of a car with a 2300 cc engine is about 42 mpg.

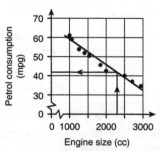

1 The table shows the engine size and the top speed of ten cars in miles per hour (mph).

| Engine size (cc) | 1600 | 1800 | 2000 | 2500 | 2900 |
|---|---|---|---|---|---|
| Top speed (mph) | 111 | 121 | 129 | 130 | 140 |

| Engine size (cc) | 1400 | 1300 | 1100 | 1000 | 2700 |
|---|---|---|---|---|---|
| Top speed (mph) | 105 | 95 | 89 | 80 | 136 |

(a) Draw a scatter graph to show this information.
(b) Draw a line of best fit on your scatter graph.
(c) Describe the correlation.

2 The outdoors temperature, in °C, at noon on ten days and the number of units of electricity used in heating a house on each of those days are shown in the table.

| Noon temperature (°C) | 7 | 11 | 9 | 2 | 4 | 7 | 0 | 10 | 5 | 3 |
|---|---|---|---|---|---|---|---|---|---|---|
| Units of electricity used | 32 | 20 | 27 | 37 | 32 | 28 | 41 | 23 | 33 | 36 |

(a) Draw a scatter graph to show this information.
(b) Draw a line of best fit on your scatter graph.
(c) Describe the correlation.
(d) Use your line of best fit to estimate the number of units of electricity used in heating a house on a day when the outdoors temperature at noon was 8°C.

3 The table shows the number of goals scored and the number of goals conceded by ten soccer teams. What can you deduce from it?

| Number of goals scored | 13 | 16 | 18 | 19 | 22 | 24 | 26 | 30 | 33 | 34 |
|---|---|---|---|---|---|---|---|---|---|---|
| Number of goals conceded | 20 | 26 | 15 | 11 | 23 | 18 | 11 | 25 | 12 | 19 |

# 6.8 Discrete and continuous data

There are two types of numerical statistical data, discrete data and continuous data.

**Discrete** data can take only exact values. Discrete data are often obtained by counting, in which case the values will be whole numbers. Examples of discrete data are the number of goals scored in an ice hockey match and the number of baked beans in a tin.

**Continuous** data are obtained by measuring, using an instrument such as a ruler or a thermometer, and cannot, therefore, be exact. Examples of continuous data are the length of a room and the weight of an apple.

### EXERCISE 6.5

1 State whether each of these types of data is discrete or continuous.
   (a) The number of brothers and sisters you have.
   (b) The time it takes you to walk a mile.
   (c) The temperature in New York at noon.
   (d) The score on a dice.
   (e) The number of seeds in a packet.
   (f) Your weight.
   (g) Your height.
   (h) Your shoe size.
   (i) The volume of wine in a bottle.
   (j) The number of words on a page in this book.

## 6.9 Grouping data

Here are the marks gained by 40 candidates in an examination.

| | | | | | | | | | |
|---|---|---|---|---|---|---|---|---|---|
| 4 | 9 | 11 | 14 | 18 | 20 | 24 | 25 | 28 | 31 |
| 32 | 32 | 35 | 38 | 39 | 40 | 41 | 41 | 43 | 44 |
| 46 | 46 | 48 | 49 | 52 | 55 | 56 | 58 | 58 | 60 |
| 65 | 67 | 69 | 74 | 75 | 78 | 80 | 83 | 86 | 94 |

A vertical line graph shows the frequency of every individual mark but it does not give a helpful picture of the results, as it shows too much detail.

Grouping the marks gives a clearer picture. The marks could be grouped 0–19, 20–39, 40–59 and so on. These are called the **class intervals**. These intervals must not overlap. You could not, for example, use class intervals of 0–20, 20–40, 40–60 etc., as you would not know where to put a mark of 20. It could go in more than one class interval.

The number of candidates with marks in each class interval is shown in the **grouped frequency table**.

| Mark | 0–19 | 20–39 | 40–59 | 60–79 | 80–99 |
|------|------|-------|-------|-------|-------|
| Frequency | 5 | 10 | 14 | 7 | 4 |

You can show this information on the bar chart shown in Figure 6.14.

You could also draw without gaps between the bars, as in Figure 6.15.

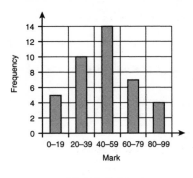

**Figure 6.14**    **Figure 6.15**

Another way to show the information is with a **frequency polygon,** obtained by joining the mid-points of the tops of the bars with straight lines, as in Figure 6.16. Figure 6.17 shows the frequency polygon.

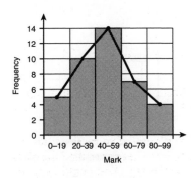 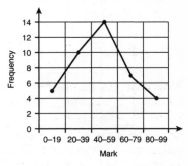

**Figure 6.16**    **Figure 6.17**

Examination marks are discrete data. Grouping continuous data is different in two ways. One of these is the way in which class intervals are defined. For example, with class intervals which have gaps, such as 60–64, 65–69 and so on, where would you put a value of 64.8?

Extending the 60–64 interval up to but not including 65 over-comes this problem. One way of writing such a class interval is $60 \leq w < 65$, where $w$ represents the quantity being measured. $60 \leq w < 65$ is read as '$w$ is greater than or equal to 60 and less than 65'.

The other difference concerns bar charts which show continuous data. The numbers on the horizontal axis are written in the usual way and there must not be any gaps between the bars.

---

**Example 6.9.1**

The table gives information about the weights, in kg, of 50 men.

| Weight ($w$ kg) | Frequency |
|---|---|
| $60 \leq w < 65$ | 4 |
| $65 \leq w < 70$ | 6 |
| $70 \leq w < 75$ | 8 |
| $75 \leq w < 80$ | 11 |
| $80 \leq w < 85$ | 15 |
| $85 \leq w < 90$ | 6 |

Show this information on
(a) a bar chart   (b) a frequency polygon.

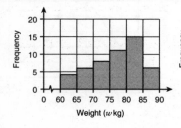
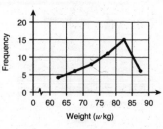

## EXERCISE 6.6

**1** Here are the numbers of days 30 people were absent from work in a month.

```
0  0  0  0  0  1  1  1  2  2  3  3  3  4  5
5  6  6  6  8  9  10 10 11 13 14 16 18 19 22
```

(a) Using groups of 0–4, 5–9 etc., show this data in a grouped frequency table.
(b) Draw a bar chart and a frequency polygon to show the data.

**2** The numbers of letters in each of the first 25 words in a book are listed below.

```
2  6  8  1  11  3  7  3  5  9  8  2  6
2  3  4  2  12  2  4  8  10  6  2  5
```

(a) Using groups of 1–3, 4–6, 7–9 and 10–12, show this information in a grouped frequency table.
(b) Draw a bar chart and a frequency polygon to show the grouped data.

**3** The grouped frequency table gives information about the times, in minutes, 25 people take to travel to work.

| Time ($t$ minutes) | Frequency |
|---|---|
| $0 \leq t < 5$ | 4 |
| $5 \leq t < 10$ | 9 |
| $10 \leq t < 15$ | 7 |
| $15 \leq t < 20$ | 3 |
| $20 \leq t < 25$ | 2 |

Use the data to draw
(a) a bar chart, (b) a frequency polygon.

**4** Here are the number of hours of sunshine recorded on each of 20 days in a town.

```
7.3  2.1  5.2  0.7  2.0  6.4  6.0  9.8  1.8  4.0
3.7  3.0  4.1  3.9  4.9  8.0  1.9  3.1  5.0  0.1
```

(a) Use the data to make a grouped frequency table using intervals

$0 \leq t < 2, 2 \leq t < 4$, etc.

(b) Draw a bar chart to show the grouped data.

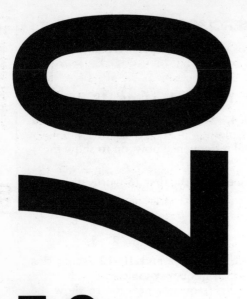

# 07

## directed numbers

**In this chapter you will learn:**

- about the meaning of negative numbers and directed numbers
- how to order directed numbers
- how to add and subtract directed numbers
- how to multiply and divide directed numbers.

# 7.1 Introduction

Numbers greater than zero are called **positive**. You can write them with a + sign in front of them but they are usually written without a sign, and assumed to be positive. Numbers less than zero are called **negative** numbers. You *must* write a – sign in front of them. Collectively, they are called **directed numbers**. Positive numbers may be represented in one direction (usually to the right or upwards on a number line), and negative numbers in the opposite direction.

Hindu mathematicians were, from the seventh century AD, the first to make a detailed study of negative numbers. It was not, however, until the Renaissance, 1000 years later, that, through the work of Fermat, Descartes and others, the concept was fully understood.

# 7.2 Ordering directed numbers

Negative numbers are used on thermometers to show temperatures below freezing, that is, below 0°C. The thermometer in Figure 7.1 shows a temperature of −10°C.

You can see from the scale that −5°C is a higher temperature than −10°C, that is, −5 > −10, and that −20°C is a lower temperature than −10°C, that is, −20 < −10.

**Figure 7.1**

A number line like that in Figure 7.2 might help you decide, for example, whether −1 is larger than −3 or not. As −1 is to the right of −3 on the number line, it is larger than −3, so −1 > −3.

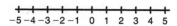

**Figure 7.2**

## EXERCISE 7.1

1  Which temperature is lower, −7°C or −5°C?
2  Which number is greater, −9 or −10?

3 Insert either < or > appropriately between these pairs of numbers.
   (a) −3, −1      (b) 0, −2      (c) −12, −15
4 Which of the temperatures −8°C, 0°C, −6°C, 2°C, 1°C is
   (a) the highest,      (b) the lowest?
5 Write these numbers in order of size, starting with the lowest.
   2, −1, 5, 0, −4, −3

## 7.3 Addition and subtraction

If a temperature of −3°C increases by 5 degrees, the new temperature is 2°C. You can express this as an addition, −3 + 5 = 2, and illustrate it on the number line in Figure 7.3. You start at −3 and go 5 to the right.

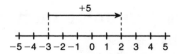

**Figure 7.3**

You can also obtain similar results such as −3 + 2 = −1 and −3 + 3 = 0. If a temperature of 2°C falls by 3 degrees, the new temperature is −1°C. You can express this as a subtraction, 2 − 3 = −1, and illustrate it on the number line in Figure 7.4. You start at 2 and go 3 to the left.

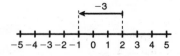

**Figure 7.4**

In a similar way, you could obtain results such as −2 − 1 = −3.

An addition like 4 + (−1) has to be written in a different form before you can apply this method. These three simple additions can be used to introduce the different form.

$$4 + 2 = 6 \qquad 4 + 1 = 5 \qquad 4 + 0 = 4$$

Continuing the pattern, 4 + (−1) = 3. However, as 4 − 1 is also 3, this suggests that adding −1 is the same as subtracting 1.

**Example 7.3.1**

Work out $-2+(-5)$.

Adding $-5$ is the same as subtracting 5,
so $-2+(-5)=-2-5=-7$.

You can devise a method for a subtraction like $5-(-1)$ using a similar approach but, in this case, you start with three simple subtractions.

$$5-2=3 \qquad 5-1=4 \qquad 5-0=5$$

Continuing the pattern, $5-(-1)=6$. However, as $5+1$ is also 6, this suggests that subtracting $-1$ is the same as adding 1.

**Example 7.3.2**

Work out $-7-(-3)$.

Subtracting $-3$ is the same as adding 3.
So $-7-(-3)=-7+3=-4$.

## EXERCISE 7.2

1 The temperature rises by 6 degrees from $-1°C$. Work out the new temperature.

2 The temperature falls by 7 degrees from $3°C$. Find the new temperature.

3 The temperature rises from $-8°C$ to $3°C$. By how many degrees does the temperature rise?

4 The temperature falls from $-2°C$ to $-9°C$. By how many degrees does the temperature fall?

5 Work out these additions and subtractions.
(a) $-3+4$ (b) $-5+2$ (c) $-4+4$ (d) $4-7$
(e) $-3-2$ (f) $-6+5$ (g) $-1-8$ (h) $-2+5-8$
(i) $-4-5+9$ (j) $7-8-3$

6 Work out these additions and subtractions.
(a) $5+(-4)$ (b) $4-(-3)$ (c) $-2+(-7)$ (d) $-6-(-5)$
(e) $-3-(-9)$ (f) $5+(-5)$ (g) $-8+(-2)$ (h) $-8-(-1)$

# 7.4 Multiplication and division

You may have used a table like the one below to find the product (see Section 1.2) of multiplying two numbers together.

| 4 | 4 | 8 | 12 | 16 |
|---|---|---|----|----|
| 3 | 3 | 6 | 9 | 12 |
| 2 | 2 | 4 | 6 | 8 |
| 1 | 1 | 2 | 3 | 4 |
| × | 1 | 2 | 3 | 4 |

You can extend the table to include negative numbers and zero. Then you can complete the rest of the table by continuing the number patterns. For example, the products in the right-hand column are 16, 12, 8 and 4, where each number is 4 less than the one above it. So the next five products will be 0, $-4$, $-8$, $-12$ and $-16$. You can complete the top left and bottom right sections of the table similarly.

The numbers in the left-hand column of the extended table are $-16$, $-12$, $-8$, $-4$ and 0 i.e. each number is 4 more than the one above it. So the next four numbers will be 4, 8, 12 and 16. You can now complete the table with positive numbers in the bottom left section.

| 4 | −16 | −12 | −8 | −4 | 0 | 4 | 8 | 12 | 16 |
|---|-----|-----|----|----|---|---|---|----|----|
| 3 | −12 | −9 | −6 | −3 | 0 | 3 | 6 | 9 | 12 |
| 2 | −8 | −6 | −4 | −2 | 0 | 2 | 4 | 6 | 8 |
| 1 | −4 | −3 | −2 | −1 | 0 | 1 | 2 | 3 | 4 |
| 0 | 0 | 0 | 0 | 0 | 0 | 0 | 0 | 0 | 0 |
| −1 | 4 | 3 | 2 | 1 | 0 | −1 | −2 | −3 | −4 |
| −2 | 8 | 6 | 4 | 2 | 0 | −2 | −4 | −6 | −8 |
| −3 | 12 | 9 | 6 | 3 | 0 | −3 | −6 | −9 | −12 |
| −4 | 16 | 12 | 8 | 4 | 0 | −4 | −8 | −12 | −16 |
| × | −4 | −3 | −2 | −1 | 0 | 1 | 2 | 3 | 4 |

Reading from the table, $(-3) \times (-2) = 6$, $3 \times (-2) = -6$ and $(-3) \times 2 = -6$.

This suggests that if two numbers have the **same** sign (both positive or both negative) their product is **positive**.

If two numbers have **different** signs, their product is **negative**.

To multiply directed numbers, use these rules to decide if the product is positive or negative; then multiply the numbers, 'ignoring' the signs.

---

**Example 7.4.1**

Work out (a) $6 \times (-5)$, (b) $(-9) \times (-4)$, (c) $(-7) \times 0$.

(a) The numbers have different signs (the 6 is the same as +6) and their product is negative. Ignoring the signs, $6 \times 5 = 30$, so $6 \times (-5) = -30$.

(b) As the numbers have the same sign, their product is positive. Ignoring the signs, $9 \times 4 = 36$, so $(-9) \times (-4) = 36$.

(c) 0 multiplied by *any* number is still 0, so $(-7) \times 0 = 0$.

---

To carry out a division such as $20 \div 4$ you need to ask 'What number do you multiply 4 by to get 20?' Then $20 \div 4 = 5$ because $4 \times 5 = 20$. The quotient (see Section 1.2) is 5.

In the same way, $(-20) \div (-4) = 5$ and $-20 \div 5 = -4$ because $5 \times (-4) = -20$. Similarly $20 \div (-4) = -5$, because $(-4) \times (-5) = 20$.

The rules for dividing directed numbers are the same as those for multiplying. If two numbers have the **same** sign, their quotient is **positive**. If the signs are **different**, their quotient is **negative**.

To divide directed numbers, use these rules to decide if the quotient is positive or negative. Then do the division, 'ignoring' the signs.

---

**Example 7.4.2**

Work out (a) $(-24) \div (-3)$, (b) $\dfrac{-28}{4}$ (c) $0 \div (-5)$.

(a) As the numbers have the same sign, the quotient is positive. Ignoring the signs, $24 \div 3 = 8$, so:
$(-24) \div (-3) = 8$.

(b) $\dfrac{-28}{4}$ is the same as $(-28) \div 4$. The numbers have different signs so the quotient is negative. Ignoring signs, $28 \div 4 = 7$, so $\dfrac{-28}{4} = -7$.

(c) 0 divided by *any* number is still 0, so $0 \div (-5) = 0$.

---

## EXERCISE 7.3

1 Use the table to write down these products.
   (a) $2 \times (-4)$    (b) $(-4) \times (-3)$    (c) $0 \times (-2)$
   (d) $(-3) \times 4$    (e) $(-4) \times (-4)$

2 Work out these multiplications.
   (a) $(-5) \times 4$    (b) $(-6) \times (-4)$    (c) $(-7) \times 3$
   (d) $(-9) \times 0$    (e) $(-6) \times (-6)$    (f) $8 \times (-8)$
   (g) $(-5) \times (-7)$    (h) $0 \times 3$    (i) $(-8) \times 9$

3 Work out these divisions.
   (a) $18 \div (-3)$    (b) $(-32) \div (-4)$    (c) $\dfrac{-21}{7}$

   (d) $\dfrac{-15}{-3}$    (e) $\dfrac{0}{-8}$    (f) $\dfrac{30}{-5}$

   (g) $(-32) \div 8$    (h) $(-36) \div (-9)$    (i) $49 \div (-7)$

# 7.5 Using a calculator

You can use a calculator to answer questions involving negative numbers, and a calculator is particularly helpful when the numbers are either large or not whole numbers. Calculators vary in the way they deal with negative numbers, but to key in a negative number, you usually have to key in the figures first and then press the $+/-$ key.

Referring to the instructions if necessary, use your calculator to answer some of the questions in earlier exercises in this chapter. When you have done this successfully, use your calculator to carry out the calculations in Exercise 7.4.

## EXERCISE 7.4

1 Carry out the following calculations.
   (a) $23 - 431$    (b) $-37 - (-93)$    (c) $-96 + (-43)$
   (d) $(-442) \div 17$    (e) $-1.7 - 3.4$    (f) $-4.3 - (-2.7)$
   (g) $3.8 \times (-0.7)$    (h) $(-2.88) \div (-0.6)$

# 08

# graphs 1

**In this chapter you will learn:**

- how to use coordinates
- how to draw and use straight line graphs
- about lines parallel to the axes
- about regions and inequalities.

# 8.1 Coordinates

Coordinates are used in mathematics to describe the position of a point. The most commonly used system of coordinates is rectangular or cartesian coordinates.

In Figure 8.1, the horizontal line is called the **x-axis** and the vertical line is called the **y-axis**. The two axes meet at the point O, which is called the **origin**. If you start from the origin and go 3 units to the right and then 2 units up, you reach the point A, which is said to have **coordinates** (3,2). The first number, 3, is the **x-coordinate** of A and the second number, 2, is the **y-coordinate** of A. The x-coordinate and the y-coordinate of the origin are both 0 so the coordinates of the origin are (0,0).

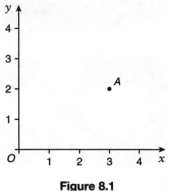

**Figure 8.1**

The concept of coordinates is ancient. Egyptian surveyors laid out towns in the form of a rectangular grid and the Greeks used the idea of coordinates to study curves. Most significantly, however, it was a coordinate system which enabled René Descartes (1596–1650) to use algebra for solving geometrical problems, thus creating analytical geometry. It is from his surname that the adjective 'cartesian' is derived.

Squared paper or graph paper is normally used for plotting points whose coordinates are given. You may omit some numbers from the scales on the axes. This saves time and avoids cluttering up the diagram. To describe the positions of points left of the y-axis or below the x-axis, you have to extend both scales below zero using negative numbers (see Figure 8.2).

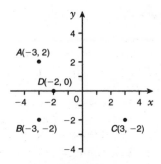

**Figure 8.2**

You can give the coordinates of a point by giving first its $x$-coordinate and then its $y$-coordinate. For example, in Figure 8.2, the $x$-coordinate of $A$ is $-3$ and its $y$-coordinate is 2. So the coordinates of $A$ are $(-3, 2)$. Similarly, the coordinates of $B$ are $(-3, -2)$, the coordinates of $C$ are $(3, -2)$ and $D$ is the point $(-2, 0)$.

## EXERCISE 8.1

1 Write down the coordinates of the points shown in the diagram.

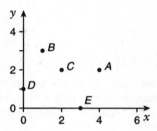

2 (a) Draw $x$- and $y$-axes from 0 to 5.
  (b) Plot and label the points $A(2, 3), B(5, 1), C(1, 0)$ and $D(0, 4)$.

3 (a) Draw $x$- and $y$-axes from 0 to 6.
  (b) Plot and label the points $E(2, 1), F(2, 4)$ and $G(6, 4)$.
  (c) $EFGH$ is a rectangle. Find the coordinates of the point $H$.

4 Write down the coordinates of the points shown in the diagram.

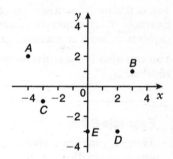

In Questions 5–8, draw $x$- and $y$-axes from $-5$ to 5.

5 Plot and label the points $A(-3, 4), B(-2, -3), C(4, -5),$ $D(0, -4)$ and $E(-4, 0)$.

6 (a) Plot and label the points $D(2, 2), E(-3, -1), F(2, -4)$ and $G(4, -1)$. Join the points in order with straight lines.
  (b) What type of quadrilateral is $DEFG$?

7 (a) Plot and label the points $P(3, 2), Q(-2, 2)$ and $R(-4, -2)$.
  (b) $PQRS$ is a parallelogram. Find the coordinates of $S$.

8 Find the coordinates of the mid-point of the line joining the points $(-5, 4)$ and $(3, 2)$.

# 8.2 Straight line graphs

You can use straight line graphs to show a relationship between two quantities. For example, the amounts of petrol used by a car in travelling certain distances are given in the table.

| Distance travelled in km | 0 | 100 | 200 | 300 |
|---|---|---|---|---|
| Petrol used in litres | 0 | 12 | 24 | 36 |

You can draw a graph to show this information by plotting points with coordinates $(0, 0)$, $(100, 12)$, $(200, 24)$ and $(300, 36)$ (see Figure 8.3).

You can read information from the graph but you have to be careful reading the scales.

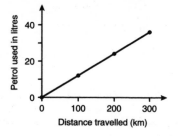

**Figure 8.3**

For example, to estimate the amount of petrol used to travel 220 km, read up to the line from 220 on the distance axis and then read across to the petrol axis. It is about 26 litres. Reverse the process to estimate the distance the car will travel on 14 litres. This is about 117 km.

You can also use straight line graphs to convert systems of units. Example 8.2.1 converts metric to Imperial units. This type of graph is called a **conversion graph.**

You can also use a conversion graph to convert one currency to another.

---

**Example 8.2.1**

The diagram is a conversion graph between kilograms (kg) and pounds (lb).

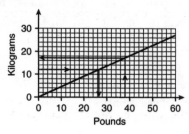

(a) Convert 38 lb to kilograms.
(b) Convert 12 kg to pounds.

(a) Read from 38 on the pounds axis up to the line and then follow the arrow across to the kilograms axis to the number 17. Thus, 38 lb is about 17 kg.

(b) Read across to the line from 12 on the kilograms axis and then follow the arrow down to the pounds axis. So 12 kg is about 26 lb.

Notice that the answers are approximate, that is, they are not exact.

In future in this book, most graphs will be drawn without the small grid of squares. This will enable you to see more clearly what is happening. For your own work, you should use graph paper.

Another type of straight line graph is the **distance-time** graph which is used to represent a journey. Use the horizontal axis for time and the vertical axis for distance. Figure 8.4 shows a distance-time graph for a cycle ride.

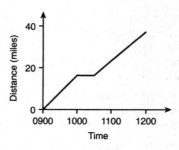

Between 0900 and 1000, the cyclist travels 19 miles.

**Figure 8.4**

Between 1000 and 1030 (the horizontal part of the graph), the cyclist is stationary, resting perhaps.

Finally, between 1030 and 1200, the cyclist travels a further 20 miles.

You can work out the **average speed** for the whole journey using

$$\text{average speed} = \frac{\text{distance travelled}}{\text{time taken}}.$$

The cyclist travelled a distance of 36 miles in 3 hours, so the average speed is $36 \div 3 = 12$. As the distance is measured in miles, and the time in hours, the speed is in miles per hour (mph).

1 The diagram shows the number of electricity units used by a 60 watt light bulb left switched on for up to 30 hours.
   (a) How many units of electricity are used in
      (i) 7 hours,
      (ii) 23 hours?
   (b) For how many hours was the bulb left on if it used
      (i) 0.7 units,
      (ii) 1.1 units?

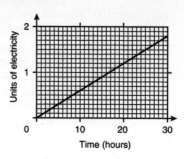

2 The graph converts miles to kilometres and vice versa.
   (a) Convert to kilometres
      (i) 28 miles,
      (ii) 42 miles.
   (b) Convert to miles
      (i) 35 km,
      (ii) 88 km.

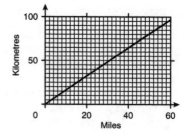

3 This graph is used to convert between pounds and euros (€).
   (a) Convert to euros (i) £38, (ii) £26.
   (b) Convert to pounds (i) €75, (ii) €48.

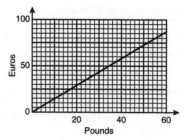

4 Mrs Ford drove from London to Brighton and back. The diagram overleaf shows the distance-time graph for her journey.

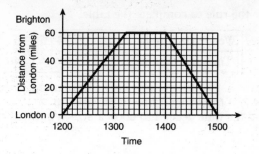

(a) At what time did she reach Brighton?
(b) For how long did she stay in Brighton?
(c) Work out her average speed for the whole journey.
(d) Work out her average speed for the journey from London to Brighton. (Hint: You do *not* work out $60 \div 1.15$.)

5 (a) Using a scale of 1 cm to 5 gallons on the horizontal axis and a scale of 1 cm to 25 litres on the vertical axis, plot the points given by the pairs of values in the table. Join the points with a straight line.

| Gallons | 0 | 10 | 20 | 30 | 40 |
|---------|---|----|----|----|----|
| Litres | 0 | 45 | 90 | 135 | 180 |

(b) Convert 18 gallons to litres.
(c) Convert 120 litres to gallons.

6 (a) Using £20 = $39 and £40 = $78, draw a graph which you can use to convert between pounds and dollars for amounts up to £50.
(b) Convert £17 to dollars.
(c) Convert $53 to pounds.

7 A cyclist set off from home at 8 a.m., and at 9.30 a.m., had covered 30 km. He rested for half an hour and then cycled a further 15 km in an hour. Finally, he cycled back home, arriving at 1 p.m.
(a) Draw a distance-time graph for the journey.
(b) Calculate his average speed for the whole journey.
(c) Calculate his average speed for the first 30 km of his journey.

8 The time, in minutes, needed to cook a chicken is given by this rule 'Multiply the weight in pounds by 20 and then add 20.'

(a) Use the rule to complete the table.

| Weight in pounds | 1 | 2 | 3 | 4 | 5 | 6 |
|---|---|---|---|---|---|---|
| Time in minutes | | 60 | | | | 140 |

(b) Draw a graph to show this information. (You should *not* join the straight line to the origin.)

(c) Find the time needed to cook a 4.2 pound chicken.

## 8.3 Lines parallel to the axes

Points with coordinates $(3, 4)$, $(3, 1)$ and $(3, -2)$ all lie on the vertical line shown in Figure 8.5.

The three points have the same $x$-coordinate, 3. In fact, *every* point on that vertical line has 3 as its $x$-coordinate and $x = 3$ is called the **equation** of the line. *Any* vertical line, therefore, has an equation of the form $x = c$, where $c$ is the value of $x$ where the line crosses the $x$-axis. The equation of the $y$-axis is $x = 0$.

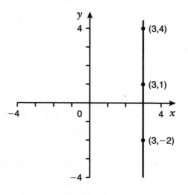

**Figure 8.5**

You can use a similar approach to find the equation of a horizontal line, except that, in this case, it will be the $y$-coordinates which are equal. So the equation of a horizontal line has the form $y = d$, where $d$ is the value of $y$ where the line crosses the $y$-axis. The equation of the $x$-axis is $y = 0$.

In Figure 8.6 the **region** to the right of the line $x = 1$ has been shaded. The $x$-coordinate of all the points in the shaded region is greater than 1.

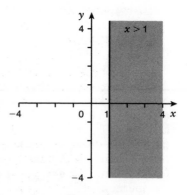

**Figure 8.6**

The line $x = 1$ is drawn as a solid line, without gaps, which tells you that the region includes the line $x = 1$. (If the line had not been part of the region, it would be dotted.) The $x$-coordinate of every point in the region is, therefore, equal to or greater than 1. The **inequality** $x \geq 1$ describes the region.

In a similar way, you can also describe regions which have one or more edges parallel to the $x$-axis using inequalities.

**Example 8.3.1**

Draw regions to illustrate the inequalities

(a) $y < 2$, (b) $-1 \leq y < 2$.

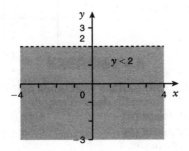

(b) Start by drawing the line $y = 2$ but it must be dotted, as points on the line $y = 2$ are not in the region. Points below the line have a $y$-coordinate less than 2, so you shade the region below the line shown in the first diagram.

(c) The inequality $-1 \leq y < 2$ is a combination of the inequalities $-1 \leq y$ and $y < 2$. As $-1 \leq y$ is the same as $y \geq -1$, the region which represents $-1 \leq y < 2$ combines the regions which represent $y \geq -1$ and $y < 2$. This is shown in the second diagram.

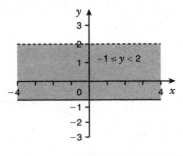

In this chapter, required regions are shown by shading but, sometimes, *unwanted* regions are shaded. Always make sure that you know which convention is being followed.

## EXERCISE 8.3

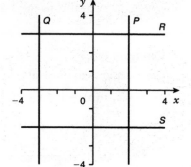

**1** Write down the equations of lines $P$, $Q$, $R$ and $S$ shown in the diagram.

**2** Draw $x$- and $y$-axes from $-5$ to $5$, and draw the following lines.
   (a) $x = -4$   (b) $y = 1$   (c) $x = 1$   (d) $y = -1$

**3** Use inequalities to describe the shaded region in each figure.

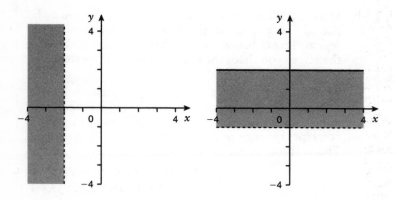

**4** Draw regions to illustrate the following inequalities.
   (a) $y \geq 2$      (b) $x \leq 0$
   (c) $-3 < x \leq 1$   (d) $-3 \leq y \leq 0$

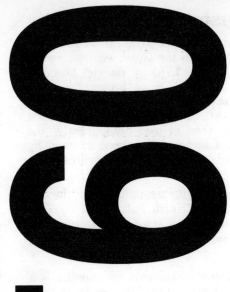

# measurement

**In this chapter you will learn:**

- about the metric system
- how to convert between metric units
- about Imperial units and how to convert between them
- how to convert between metric and Imperial units
- how to choose suitable units.

# 9.1 The metric system

The metric system is a system of units based on the **metre** (*meter* in the USA), which is derived from the Greek *metron,* meaning 'measure'. It was proposed in 1670 by Gabriel Mouton, a priest in Lyons. In the 1790s, France became the first country to adopt the metric system when the revolutionary government introduced it to bring order to the chaotic diversity of traditional units then in use. Subsequently, the metric system of weights and measures has been adopted by the majority of countries and is used by scientists throughout the world.

A metre was originally defined as one ten-millionth of the distance around the Earth from the North Pole to the equator. It has been redefined several times since and is currently defined as the distance travelled by light in $\frac{1}{299\,792\,458}$ of a second.

The symbol 'm' after a number means that the number is a measurement in metres. Thus 100 m means 100 metres.

Latin prefixes are used to express fractions of a metre. The prefix 'milli' means $\frac{1}{1000}$. Thus, 1 millimetre $= \frac{1}{1000}$ metre. The short form for a millimetre is mm so $1\,mm = \frac{1}{1000}\,m$. Thus $1\,m = 1000\,mm$.

The prefix 'centi' means $\frac{1}{100}$. Thus 1 centimetre $= \frac{1}{100}$ metre so 1 metre $= 100$ centimetres. The short form of centimetre is cm, so $1\,cm = \frac{1}{100}\,m$, and $1\,m = 100\,cm$.

Since $1\,m = 1000\,mm$ and $1\,m = 100\,cm$, $10\,mm = 1\,cm$.

Greek prefixes are used to express multiples of a metre. For example, 1 kilometre, written 1 km, is 1000 metres, so $1\,km = 1000\,m$.

These prefixes are also used with other metric units.

The main metric units of mass, often called weight, are **grams** (g) **milligrams** (mg) and **kilograms** (kg). For large masses, the **tonne** (t) ($= 1000\,kg$) is used.

The main metric units of capacity, which is the amount a container can hold, are **litres** (l) and **millilitres** (ml).

To convert from one metric unit to another, you multiply or divide by 10 or 100 or 1000, as appropriate. To convert to a *larger* unit, you *divide* and, to convert to a *smaller* unit, you *multiply*.

**Example 9.1.1**

Convert the following, using appropriate abbreviations for the units.

(a) 38 millimetres to centimetres

(b) 2.41 metres to centimetres

(c) 8753 grams to kilograms

(d) 5.7 litres to millilitres

(a) 38 mm = 3.8 cm

(b) 2.41 m = 241 cm

(c) 8753 g = 8.753 kg

(d) 5.7 l = 5700 ml

## EXERCISE 9.1

1 Convert these quantities.

(a) 79 millimetres to centimetres

(b) 850 centimetres to metres

(c) 3240 grams to kilograms

(d) 4125 millilitres to litres

(e) 1.8 kilometres to metres

(f) 400 metres to kilometres

(g) 9.4 centimetres to millimetres

(h) 0.94 litres to millilitres

(i) 2.3 kilograms to grams

(j) 3.72 tonnes to kilograms

2 Measure the length of this line. Write your answer in
(a) centimetres     (b) millimetres.

3 A dishwasher uses 25 g of powder for each wash. For how many washes will a 3 kg box of powder last?

4 A paper towel is 242 mm long. There are 230 paper towels on a roll. Find the length, in metres, of the roll.

5 At birth, a baby weighed 3.29 kg. She gained 160 g each week. Find her weight, in kilograms, after 13 weeks.

6 Claire takes 3 teaspoons of medicine twice a day. A teaspoon holds 5 ml. How many days will 1 litre of medicine last?

# 9.2 Imperial units

English weights and measures, usually called **Imperial units**, are gradually being superseded by metric units. Indeed, at the time of writing (2008), the only Imperial units which still have legal status in the United Kingdom are miles and pints, although several others are still in everyday use. Here are the most common ones and the relationships between them.

| Length | 12 inches (in) = 1 foot (ft) |
|---|---|
| | 3 feet (ft) = 1 yard (yd) |
| | 1760 yards (yd) = 1 mile |
| Mass | 16 ounces (oz) = 1 pound (lb) |
| Capacity | 8 pints (pt) = 1 gallon |

Some Imperial units have interesting backgrounds. If you stretch your arm out sideways, you will find that the distance from the end of your nose to the tip of your fingers is almost exactly a yard, a fact used by King Henry I when he fixed this measurement, taken from his own body, as the standard yard.

'Mile' comes from the Latin *mille*, meaning a thousand. A mile was the distance covered in a thousand paces, a pace being *two* strides.

Over the centuries, the British made adjustments to Imperial units to correct inconsistencies but these changes were not always implemented in their colonies. Consequently, colonial weights and measures were sometimes slightly different from those in Britain. For example, a British gallon is equal to 1.201 US gallons.

### EXERCISE 9.2

Express the given quantity in the new unit.

1. (a) 8 feet to inches
   (b) 40 pints to gallons
   (c) 48 ounces to pounds
   (d) 7 gallons to pints
   (e) 24 feet to yards
   (f) 3 miles to yards
   (g) 5 ft 10 in to inches
   (h) 3 lb 7 oz to ounces
   (i) 43 inches to feet and inches
   (j) 57 oz to pounds and ounces

2. Paul is 5 ft 10 in tall. Peter is 4 inches taller than Paul. Find Peter's height in feet and inches.

3. Bill is 6 ft 3 in tall and Ann is 5 ft 7 in. How many inches taller is Bill than Ann?

4. Margaret drinks a pint of milk every day. How many gallons of milk will she drink in 16 weeks?

5. The biggest beer tankard in the world holds 615 gallons. How many pints is this?

6. My stride is 32 inches long. How many strides do I make in walking 1 mile?

# 9.3 Converting between metric and Imperial units

It is useful to be able to convert between metric and Imperial units in common use. For most practical purposes, approximate equivalents are adequate. Here are the main ones.

**Length**  1 inch = 2.5 centimetres
     1 metre = 3 feet
     1 mile = 1.6 kilometres
   i.e. 5 miles = 8 kilometres
**Mass**  1 kilogram = 2.2 pounds
     1 ounce = 25 grams
**Capacity**  1 litre = $1\frac{3}{4}$ pints
     1 gallon = 4.5 litres

---

**Example 9.3.1**

(a) Convert 15 centimetres to inches.
(b) Convert 30 miles to kilometres.

(a) There are 2.5 cm in 1 inch. In 15 cm there are $15 \div 2.5$ in = 6 in.
(b) One mile is 1.6 km, so 30 miles is $30 \times 1.6$ km = 48 km.
  Alternatively, you could use 5 miles = 8 km.
  You must multiply 5 by 6 to get 30, so 30 miles = 6 × 8 km = 48 km.

---

## EXERCISE 9.3

Convert the first quantity to the second.

1 (a) 6 inches to centimetres (b) 35 centimetres to inches
 (c) 4 metres to feet (d) 45 miles to kilometres
 (e) 150 grams to ounces (f) 5 kilograms to pounds
 (g) 12 litres to pints (h) 54 litres to gallons
 (i) 96 kilometres to miles (j) 14 pints to litres

2 The waist size of a pair of trousers is 32 inches. How many centimetres is this?

3 A tin contains 425 g of soup. How many ounces is this?

4 The chest size of a sweater is 105 cm. How many inches is this?

5 The capacity of a car's petrol tank is 36 litres. How many gallons is this?

6 David weighs 85 kg. Express his weight in pounds.

7 The distance from London to Birmingham is 120 miles. Express this in kilometres.

8 Barbara's height is 175 cm. Express this in feet and inches.

9 The distance from London to Cardiff is 248 km. Express this in miles.

10 Delia needs 10 lb of sugar to make jam. Sugar is sold in 1 kg bags. How many bags must she buy?

11 A car travels 45 miles on one gallon of petrol. How many kilometres will it travel on one litre of petrol?

## 9.4 Choosing suitable units

Before you measure something, you must decide which units to use. To measure the width of this page in metric units, for example, centimetres or millimetres are the most appropriate. Expressed in these units, the width is 12.7 cm or 127 mm. If you used metres or kilometres, you would need numbers less than 1 for your answers.

On the other hand, the most sensible metric units for the weight of a car are kilograms or tonnes. 2200 kg or 2.2 tonnes are reasonable ways of expressing the weight of a Ford Mondeo but, if you used grams or milligrams, you would require very large numbers for your answers.

### EXERCISE 9.4

1 Write down the name of
  (i) an appropriate metric unit and
  (ii) an appropriate Imperial unit for measuring each of the following.
  (a) the height of the Eiffel Tower
  (b) the weight of a coin
  (c) the amount of water in a kettle
  (d) the radius of the Earth
  (e) the amount of water in a pond
  (f) the weight of a television

2 Complete each if these sentences by writing an appropriate metric unit to make the statement true.
  (a) A tin of soup weighs 425 ..........
  (b) A large bottle of lemonade holds 3 ..........
  (c) A ball of wool weighs 25 ..........
  (d) There are about 5 .......... of blood in the human body
  (e) A teaspoon holds about 5 ..........

# perimeter and area

**In this chapter you will learn:**

- how to find and use the perimeter of a rectangle
- about the meaning of area
- how to find and use the area of rectangles, triangles, parallelograms and trapezia.

# 10.1 Perimeter

The **perimeter** of a shape is the length of its boundary. The word 'perimeter' is derived from the Greek *peri* and *metron*, meaning 'around' and 'measure'.

For example, to find the perimeter of a triangle with sides of length 2 cm, 3 cm and 4 cm you add the three lengths. So the perimeter of this triangle is $(2 + 3 + 4)$ cm = 9 cm.

To find the perimeter of a square with sides of 3 cm, you could work out the sum $3 + 3 + 3 + 3$ but it is quicker to find the product $(4 \times 3)$ cm = 12 cm. Similarly, to find the perimeter of a rectangle 6 cm long and 5 cm wide, you could work out the sum $6 + 5 + 6 + 5$. Alternatively, you could double both the length and width and then add the two results, giving a perimeter of $(2 \times 6 + 2 \times 5)$ cm = $(12 + 10)$ cm = 22 cm.

In general,

**perimeter of a rectangle = 2 × length + 2 × width.**

## EXERCISE 10.1

1 The lengths of the sides of a triangle are 8 cm, 5 cm and 4 cm. Find its perimeter.

2 The length of each side of a triangle is 9 cm. Find its perimeter.

3 Find the perimeter of a square with sides 9.3 cm long.

4 A rectangle is 6.7 cm long and 4.8 cm wide. Find its perimeter.

5 Find the perimeter of a regular pentagon with sides 7.2 cm long.

6 For international matches, the length of a soccer pitch must be between 100 m and 110 m. Its width must be between 64 m and 73 m. Find the perimeter which is
(a) the least possible    (b) the greatest possible.

7 On separate diagrams, plot the following points and join them in the order given. Find the perimeter of each shape.
(a) $(1, 1), (1, 4), (3, 4), (3, 3), (5, 3), (5, 2), (4, 2), (4, 1)$.
(b) $(1, 2), (1, 3), (3, 3), (3, 5), (5, 5), (5, 1), (4, 1), (4, 2)$.

8 A soccer pitch is 100 m long and 75 m wide. A man runs around the pitch, travelling 7 km. How many times did he run round?

9 The table shows the number of rolls of wallpaper needed for a room, if the perimeter and the height of the room are known.

| Height (metres) | Perimeter not over | | | | | | | |
|---|---|---|---|---|---|---|---|---|
| | 10 m | 12 m | 14 m | 16 m | 18 m | 20 m | 22 m | 24 m |
| 2.15–2.30 | 5 | 5 | 6 | 7 | 8 | 9 | 10 | 10 |
| 2.30–2.45 | 5 | 6 | 7 | 8 | 8 | 9 | 10 | 11 |
| 2.45–2.60 | 5 | 6 | 7 | 8 | 9 | 10 | 11 | 12 |
| 2.60–2.75 | 5 | 6 | 7 | 8 | 9 | 10 | 11 | 12 |
| 2.75–2.90 | 6 | 7 | 8 | 9 | 10 | 11 | 12 | 13 |
| 2.90–3.05 | 6 | 7 | 8 | 9 | 10 | 11 | 12 | 13 |
| 3.05–3.20 | 7 | 8 | 9 | 10 | 12 | 13 | 14 | 15 |

Find the number of rolls of wallpaper needed for a room with
(a) a height of 2.40 m and a perimeter of 21 m,
(b) a height of 2.65 m and a perimeter of 18 m,
(c) a height of 2.53 m and a perimeter of 16.15 m.

**10** Use the table in Question 9 to find the number of rolls of wallpaper needed for a rectangular room which is
(a) 5 m long, 4.5 m wide and 2.84 m high,
(b) 6.24 m long, 4.86 m wide and 2.95 m high.

**11** The figure, which is not to scale, shows the plan of a room. The height of the room is 2.5 m. How many rolls of wallpaper are needed for it?

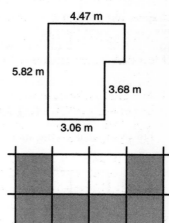

4.47 m

5.82 m

3.68 m

3.06 m

**12** The figure shows six squares on a centimetre grid measuring 4 cm by 2 cm.
(a) Find its perimeter.
(b) Draw as many more arrangements of six squares on this grid as you can; find the perimeter of each one.
(c) What is the greatest perimeter you can get?

# 10.2 Area

Area is a measurement of amount of surface. To find the area of a surface, you need to cover it with squares without leaving any gaps, and then count the number of squares used. A square with sides 1 cm long has an area of 1 square centimetre, which is written as $1\,\text{cm}^2$.

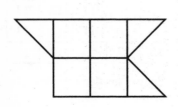

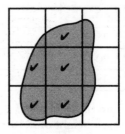

**Figure 10.1**                    **Figure 10.2**

For example, the shape shown in Figure 10.1 contains 4 full squares and 3 half squares. So its area is $\left(4 + 1\frac{1}{2}\right)\text{cm}^2 = 5\frac{1}{2}\,\text{cm}^2$.

You cannot always find the area of a shape exactly. For awkward shapes, like that in Figure 10.2, you have to approximate. One way is to count the number of complete squares inside the shape (one in this case) and those which are half or more inside (four). If less than half a square is covered by the shape, ignore it.

So the approximate area is $5\,\text{cm}^2$.

## EXERCISE 10.2

1 Find the areas of the shapes shown in parts (a) to (f).

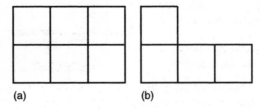

(a)                    (b)

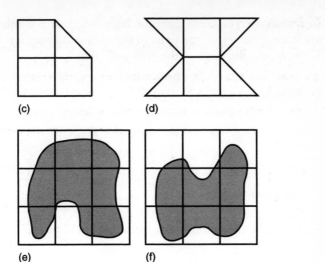

(c)            (d)

(e)            (f)

For Questions 2 and 3 use centimetre squared paper.

2 Plot the following points and join them in the order given. Find the area of each shape.
   (a) (1,1), (1,3), (4,3), (4,1)     (b) (1,1), (3,3), (6,3), (4,1)
   (c) (1,2), (4,5), (7,2)            (d) (2,1), (4,5), (7,2)
   (e) (1,2), (2,5), (6,6), (5,1)     (f) (2,1), (2,5), (3,5), (5,3)

3 Draw a circle with a radius of 3 cm. (Set your compasses to 3 cm.) Find its approximate area.

# 10.3 Area of a rectangle

The rectangle in Figure 10.3 is 3 cm long and 2 cm wide. It contains 6 centimetre squares, so its area is 6 cm². But you could count the squares differently, by saying that there are two rows of three squares, that is 3 × 2 = 6.

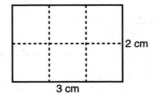

**Figure 10.3**

This is equivalent to multiplying the length of the rectangle by its width. Thus the area of the rectangle is (3 × 2) cm² = 6 cm².

In general:
   **Area of a rectangle = length × width.**

You must use the same units for both length and width: if they are both centimetres, the units of the area will be cm$^2$; if they are both metres, the units of the area will be m$^2$.

A **square** is a rectangle whose length and width are equal.

So the formula for area becomes:
> Area of a square = length of side × length of side.

You can sometimes find the area of a shape by dividing it into rectangles.

---

### Example 10.3.1

Find the area of the shape in the figure. The dotted lines divide the shape into three rectangles *A*, *B* and *C*.

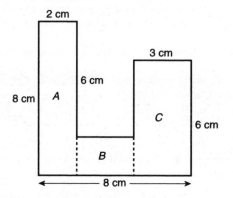

Area $A = (8 \times 2)$ cm$^2 = 16$ cm$^2$.
Area $B = (3 \times 2)$ cm$^2 = 6$ cm$^2$.
Area $C = (6 \times 3)$ cm$^2 = 18$ cm$^2$.
By adding, the total area is 40 cm$^2$.

---

## EXERCISE 10.3

1 A rectangle is 12 cm long and 7 cm wide. Find its area.

2 A rectangle is 8 m long and 6 m wide. Find its area.

3 Each side of a square is 9 cm long. Find its area.

4 A rectangle is 2.3 m long and 60 cm wide. Find its area
   (a) in m$^2$,   (b) in cm$^2$.

5 Each side of a square is 100 cm long. Find its area
(a) in m², (b) in cm².

6 Find the areas of the shapes shown in the diagrams. The lengths are in centimetres, and the diagrams are not drawn to scale.

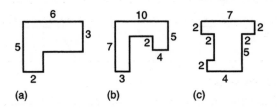

(a)  (b)  (c)

7 Find the areas of the shaded regions in the diagrams. The lengths are in centimetres, and the diagrams are not drawn to scale.

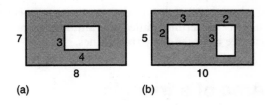

(a)  (b)

8 A room is 3 m high and its floor is a rectangle 5 m by 4 m.
(a) Find the total area of the walls.
(b) Find the area of the ceiling,
(c) One litre of paint covers 12 m². How many litres of paint are needed for the walls and the ceiling?

9 A carpet which is 3 m by 2 m is placed on a floor which is 5 m by 3 m. What area of the floor is not covered by the carpet?

10 A rectangular lawn is 9 m by 7 m. There is a path, which is 1 m wide all around the outside of the lawn. Find the area of the path.

11 The length of each side of a square is 100 m. The square has an area of 1 hectare. Find its area in m².

12 The floor of a room is a rectangle 3.5 m by 3 m. How many square carpet tiles, each 50 cm by 50 cm, are needed to cover the floor?

## 10.4 Area of a parallelogram

To find the area of the parallelogram in Figure 10.4, remove the shaded triangle and replace it at the other end to make a rectangle. This rectangle must have the same area as the original parallelogram. So

area = (3 × 2) cm² = 6 cm².

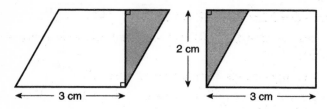

**Figure 10.4**

In general,

**area of parallelogram = base × height.**

Remember that the 'height' is the vertical (or perpendicular) height, which is sometimes called the **altitude**. In other words, it is the distance between the parallel lines.

## 10.5 Area of a triangle

Starting with any triangle, you can construct a parallelogram which has twice the area of the original triangle (see Figure 10.5). So

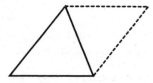

**Figure 10.5**

**area of triangle = $\frac{1}{2}$ × base × height.**

Again, remember that 'height' means *vertical* height.

To find the area of triangle $ABC$ in Figure 10.6, for example, measure the length of the base $AB$ (3 cm) and measure the length of $CD$, the perpendicular distance from $C$ on to $AB$ (1 cm).

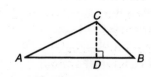

**Figure 10.6**

Area of triangle $ABC = (\frac{1}{2} \times 3 \times 1)$ cm² = $1\frac{1}{2}$ cm².

Alternatively, you could use AC or BC as the 'base'. If you used AC as the base, you would have to extend AC, before you could draw and measure BE, the 'vertical' height from B to AC (see Figure 10.7).

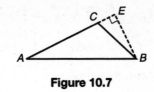

**Figure 10.7**

Then $AC = 2.2$ cm and $BE = 1.35$ cm.

So area of triangle $ABC \approx \frac{1}{2} \times 2.2 \times 1.35$ cm$^2 = 1.485$ cm$^2 \approx 1.5$ cm$^2$.

The symbol $\approx$ means 'is approximately equal to'.

You can sometimes find the area of a shape by dividing it into rectangles and triangles, or by 'boxing' it.

---

**Example 10.5.1**

Find the area of the trapezium in the figure.

The dotted line divides the trapezium into a rectangle $A$ and a triangle $B$.

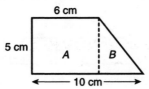

Area of rectangle $A = (6 \times 5)$ cm$^2 = 30$ cm$^2$.

Area of triangle $B = \left( \frac{1}{2} \times 4 \times 5 \right)$ cm$^2 = 10$ cm$^2$.

So total area $= 40$ cm$^2$.

---

**Example 10.5.2**

Find the area of the shaded triangle.

Area of rectangle $(8 \times 6)$ cm$^2 = 48$ cm$^2$.

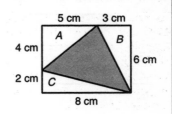

Area of triangle $A = \left(\frac{1}{2} \times 4 \times 5\right)$ cm$^2$ = 10 cm$^2$.

Area of triangle $B = \left(\frac{1}{2} \times 6 \times 3\right)$ cm$^2$ = 9 cm$^2$.

Area of triangle $C = \left(\frac{1}{2} \times 8 \times 2\right)$ cm$^2$ = 8 cm$^2$.

Sum of areas of triangles $A$, $B$ and $C = (10 + 9 + 8)$ cm$^2$
= 27 cm$^2$.

Area of shaded triangle $= (48 - 27)$ cm$^2$ = 21 cm$^2$.

## EXERCISE 10.4

1 Find the areas of the following shapes. The units are centimetres.

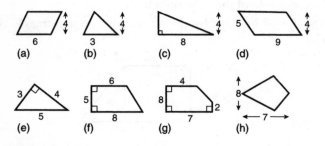

(a)     (b)     (c)     (d)

(e)     (f)     (g)     (h)

2 Make appropriate measurements to find the areas of these shapes.

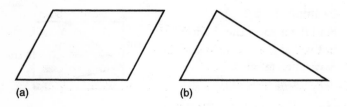

(a)                    (b)

3 Find the shaded areas. The units are centimetres.

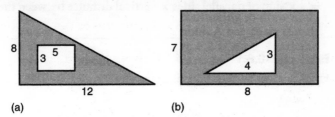

(a)  (b)

4 Using the method of Example 10.5.2, find the area of the triangle with its vertices at (0,1), (3,3) and (4,0).

5 The diagonals of a rhombus are 10 cm and 8 cm long. Find its area. (Hint: the diagonals of a rhombus cross at right angles.)

6 The diagonals of a kite are 12 cm and 10 cm long. Find its area.

# 10.6 Area of a trapezium

To find the area of the trapezium shown in Figure 10.8, join it to a congruent trapezium to form a parallelogram. (See Figure 10.9.)

The length of the base of the parallelogram is
    $(12 + 8)$ cm $= 20$ cm.

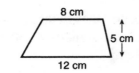

**Figure 10.8**

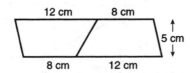

**Figure 10.9**

So the area of the parallelogram is
    $(20 \times 5)$ cm$^2 = 100$ cm$^2$.

The area of the trapezium is half that of the parallelogram, that is
    $\left(\frac{1}{2} \times 20 \times 5\right)$ cm$^2 = 50$ cm$^2$.

**Area of a trapezium**
   $= \frac{1}{2} \times$ sum of parallel sides $\times$ vertical distance between them.

---

**Example 10.6.1**

Find the area of the trapezium in the figure.

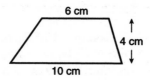

$$\text{Area} = \frac{1}{2} \times (10 + 6) \times 4 \, \text{cm}^2$$
$$= 32 \, \text{cm}^2.$$

---

You can sometimes find the area of a shape by dividing it into trapezia.

---

**Example 10.6.2**

Find the area of the pentagon in the figure.

Divide the pentagon into trapezia $A$ and $B$. Then, area of
$A = \frac{1}{2} \times (3 + 9) \times 4 \, \text{cm}^2$
   $= 24 \, \text{cm}^2.$
Area of $B = \frac{1}{2} \times (9 + 5) \times 8 \, \text{cm}^2 = 56 \, \text{cm}^2.$

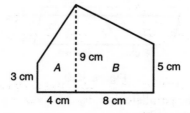

Area of pentagon $= (24 + 56) \, \text{cm}^2 = 80 \, \text{cm}^2.$

---

## EXERCISE 10.5

1 Find the area of each of the trapezia shown. The lengths are in centimetres.

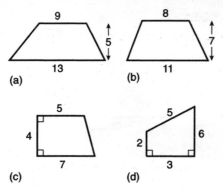

(a)   (b)

(c)   (d)

2 By taking measurements, find the area of this trapezium.

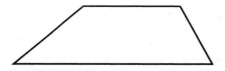

3 Find the areas of trapezia with vertices at
(a) $(1, 1), (2, 3), (5, 3), (8.1)$,   (b) $(2, 2), (2, 4), (5, 5), (5.0)$,
(c) $(2, 3), (2, 6), (6, 6), (6, 1)$.

4 Find the area of these polygons. The units are centimetres.

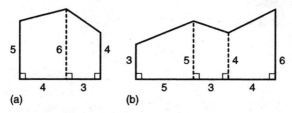

(a)   (b)

5 Two of the vertices of an isosceles trapezium are at $(1,1)$ and $(3,2)$. The line with equation $y = 3$ is a line of symmetry. Find the area of the trapezium.

6 Three of the vertices of a hexagon are at $(1,1)$, $(3,2)$ and $(1,3)$. The line with equation $x = 4$ is a line of symmetry. Find the area of the hexagon.

# algebraic expressions

**In this chapter you will learn:**

- how to write and simplify algebraic expressions
- how to evaluate algebraic expressions
- how to use brackets
- about indices
- how to use the laws of indices.

# 11.1 Introduction – what is algebra?

The Arabic word *algebra* originally meant the study of equations but has come to mean the whole branch of mathematics in which letters and symbols are used to generalize mathematical relations.

Many of the techniques of solving equations were known to the ancient Babylonians about 4000 years ago but modern, symbolic algebraic notation was not invented until the seventeenth century.

# 11.2 Writing expressions

Suppose that you are driving along the road, and that you are going at 50 mph. In one hour you will go 50 miles, and in 3 hours you will go 150 miles. To find the distance you travel, you multiply the speed by the time. So if you travel for $t$ hours, then you would travel $50 \times t$ miles. The letter $t$ stands for a number. Using the expression $50 \times t$, you can calculate the distance travelled for any time.

As letters stand for numbers, you will want to add, subtract, multiply and divide them. You can do all these processes with letters but, sometimes, the answers are written in shortened forms.

If $y$ stands for a number and you were asked to write down a number which is 3 more than $y$, think how you would answer the question if it were asked with numbers instead of letters.

Just as $4 + 3$ is 3 more than 4, so $y + 3$ (or $3 + y$) is 3 more than $y$. Similarly, the number which is 3 less than $y$ is $y - 3$. This is *not* the same as $3 - y$, which is the number that is $y$ less than 3.

The number which is 5 times greater than $y$ could be written as $5 \times y$ (or $y \times 5$), but the multiplication sign is usually omitted and this would be written as $5y$. Notice that the number is written in front of the letter. In the same way, the distance travelled by car earlier in the section would be written as $50t$ miles.

The number which is $\frac{1}{3}$ of $y$ is written $\frac{y}{3}$ or $\frac{1}{3}y$. It is not usual, but it is not wrong, to have division signs in algebraic expressions.

## Summary

$3a$ means $3 \times a$                      $ab$ means $a \times b$

$3ab$ means $3 \times a \times b$

$\frac{a}{3}$ or $\frac{1}{3}a$ means $a \div 3$          $\frac{a}{b}$ means $a \div b$

## Example 11.2.1

The sum of two numbers is 9. One of the numbers is $x$. Write down an expression for the other number.

The other number must be $x$ less than 9, which is $9 - x$.

## Example 11.2.2

A man drives at $s$ miles per hour for $t$ hours. Write down an expression for the distance he travels.

To find the distance he travels, multiply his speed by the time that he drives. The distance he travels is $s \times t$ miles, which is written as $st$ miles.

Notice that $s \times t$ is acceptable as an answer but you would normally omit the multiplication sign and write the answer as $st$ miles.

## EXERCISE 11.1

In each case, write your answer in the form suggested in the summary, without multiplication or division signs.

1 $x$ is a number. Write down an expression for a number which is
   (a) 2 more than $x$    (b) 1 less than $x$
   (c) 4 times $x$    (d) $\frac{1}{2}$ of $x$.

2 $p$ and $q$ are two numbers. Write down an expression for
   (a) the sum of $p$ and $q$    (b) the product of $p$ and $q$.

3 The product of two numbers is 8. One of the numbers is $x$. Write down an expression for the other number.

4 $a$ and $b$ are two numbers and $a$ is greater than $b$. Write down an expression for the difference between the numbers.

5 There are 30 students in a class and $b$ of them are boys. Write down an expression for the number of girls in the class.

6 Write these expressions in their simplest forms.
   (a) $6 \times x$   (b) $y \div 5$   (c) $x \times y$   (d) $3 \times x \times y$
   (e) $x \div y$   (f) $5 \times x \div y$   (g) $x \times y \div 5$   (h) $x \times y \times z$

7 Write down an expression for the number of days in $w$ weeks.

8 Write down the number of metres in $d$ centimetres.

9   A railway carriage has $p$ seats. Write down an expression for the number of seats in $n$ carriages.

10  When $a$ apples are shared amongst $b$ children, none are left over. Write down the number of apples each child receives.

# 11.3  Simplifying expressions

It is useful to be able to make algebraic expressions simpler if possible. For example:

$a + a + a + a$ is written as $4a$, so $a + a + a + a = 4a$;
$3b + 2b$ is written as $5b$, so $3b + 2b = 5b$;
$7c - 4c$ is written as $3c$, so $7c - 4c = 3c$.

However, in the expression $3d + 5e + 4d - 2e$, the $d$'s and the $e$'s must be dealt with separately.

As $3d + 4d = 7d$ and $5e - 2e = 3e$, $3d + 5e + 4d - 2e = 7d + 3e$.

Similarly, $5a - 2b - 3a + b = 2a - b$.

Notice that $1b = b$, that is, it is conventional to leave out the 1.

In the expression $7d + 3e$, $7d$ and $3e$ are called **terms**. Terms such as $3b$ and $2b$ are called **like** terms, and terms such as $7d$ and $3e$ are called **unlike** terms.

Thus, in the expression $2a + 3b + 4a - 2c$, there are four terms. Two of them, $2a$ and $4a$, are like each other, and the other terms are unlike each other, and unlike the terms $2a$ and $4a$.

Like terms can be combined by adding and subtracting them. This process is called **collecting like terms**. When the like terms have been collected, the expression is said to be **simplified**.

Unlike terms cannot be collected together in the same way.

---

**Example 11.3.1**

Simplify   (a) $6f + 5 + 2f - 3$,   (b) $3ab + 2bc - ab - abc$.

(a) Collecting like terms gives $6f + 2f = 8f$ and $5 - 3 = 2$, so $6f + 5 + 2f - 3 = 8f + 2$. You will usually write this straight down without intermediate steps.

(b) The only like terms are $3ab$ and $- ab$ so
$3ab + 2bc - ab - abc = 2ab + 2bc - abc$.

---

**Example 11.3.2**

The length of each side of an equilateral triangle is $d$ centimetres. Write down, in its simplest form, an expression for its perimeter.

The perimeter is $d + d + d$, but it is not in its simplest form, which is $d + d + d = 3d$.

## EXERCISE 11.2

1 The length of each side of a regular pentagon is $l$ centimetres. Write down and simplify an expression for its perimeter.

2 The length of a rectangle is $a$ centimetres and its width is $b$ centimetres. Write down an expression, in its simplest form, for
(a) its perimeter,     (b) its area.

3 Write down an expression for the perimeter of the hexagon in the figure, giving your answer in its simplest form.

4 The lengths, in centimetres, of the sides of a trapezium are $a, a, 2a$ and $b$. Find and simplify an expression for its perimeter.

5 Write each of these expressions in its simplest form.
   (a) $x + x$
   (b) $x + x + x + x$
   (c) $x + x + x + x + x + x + x$
   (d) $x + x - x + x + x - x + x$
   (e) $4x + 2x + 3x$
   (f) $3y + 2y - 4y$
   (g) $4xy - 7xy + xy$
   (h) $3abc + 2abc - 6abc$
   (i) $5x + 3y + 4x + 2y$
   (j) $4x + 7 - 2x - 4$
   (k) $6x + 3 - x + 2x$
   (l) $3x + 4y + 2x - 3y - y$
   (m) $7xy + 5x - 4x + 2xy - 3$
   (n) $4a + 5b + 3a + 6 - 7b$
   (o) $a + 3ab + 6 + 7a - 2ab$
   (p) $3a + 6 + 2a - 4 + a - 2$
   (q) $8 + 3pq - 5 - 7pq - 1 + r$
   (r) $5p + q - 3r + 2r - 4p + r$

# 11.4 Evaluating expressions

If you know the values of the letter or letters used in an expression, you can calculate its value. For example, if $x = 2$, then $3x - 1 = 6 - 1 = 5$. Substituting numbers for letters and calculating the value of an expression is called **evaluating** the expression.

The rules for the order in which the operations must be carried out were introduced in Chapter 01, Section 1.3. Multiplication and division are carried out before addition and subtraction.

---

**Example 11.4.1**

If $x = 5$, evaluate (a) $4x + 1$, (b) $6 - 2x$, (c) $\dfrac{x + 1}{2}$.

(a) $4x + 1 = 4 \times 5 + 1$    Multiplication comes before
           $= 20 + 1$         addition
           $= 21$.

(b) $6 - 2x = 6 - 2 \times 5$    Multiplication comes before
           $= 6 - 10$        subtraction
           $= -4$.

(c) $\dfrac{x + 1}{2} = \dfrac{5 + 1}{2} = \dfrac{6}{2}$    Evaluate the numerator before
        $= 3$.            dividing

---

You may want to leave out some steps as you become confident.

---

**Example 11.4.2**

If $x = 6$ and $y = 3$, evaluate   (a) $2xy$,   (b) $5x - 7y$,   (c) $\dfrac{x}{y}$.

(a) $2xy = 2 \times 6 \times 3 = 36$

(b) $5x - 7y = 5 \times 6 - 7 \times 3 = 30 - 21 = 9$

(c) $\dfrac{x}{y} = \dfrac{6}{3} = 2$.

---

## EXERCISE 11.3

1 If $x = 3$, evaluate the following expressions.
   (a) $5x + 4$     (b) $2x - 1$     (c) $10 - 3x$     (d) $1 - 2x$

2 If $x = 4$ and $y = 7$, evaluate the following expressions.
   (a) $2x + y$    (b) $5x - 2y$    (c) $4x + 5y$    (d) $3x - 4y$

3 If $a = 2$, $b = 3$ and $c = 10$, evaluate the following expressions.
   (a) $ab + c$    (b) $abc$    (c) $5abc$    (d) $2a + bc$

# 11.5 Squaring

The notation $x^2$ is short for $x \times x$ and is read as '$x$ squared'.

The notation $4x^2$ means $4 \times x \times x$. (Only the $x$ is squared, not the 4.)

---

**Example 11.5.1**

If $x = 5$, evaluate     (a) $x^2$,     (b) $3x^2$.

(a) $x^2 = 5 \times 5 = 25$.     (b) $3x^2 = 3 \times 5 \times 5 = 3 \times 25 = 75$.

---

**Example 11.5.2**

If $x = -3$, evaluate $6x^2 - 7$.

$6x^2 - 7 = 6 \times (-3)^2 - 7 = 6 \times 9 - 7 = 54 - 7 = 47$.

---

### EXERCISE 11.4

1   If $x = 3$, evaluate the following.
    (a) $x^2$     (b) $x^2 + 1$     (c) $5x^2$     (d) $2x^2 - 21$

2   If $y = -4$, evaluate the following.
    (a) $y^2$     (b) $y^2 - 3$     (c) $2y^2 + 5$     (d) $2y^2 - 2y$

3   If $a = 5$, $b = -2$, $c = 6$ and $d = 2$ evaluate the following.
    (a) $4a^2 - 3a$   (b) $3b^2 - 5b$   (c) $2c^2 + 4$   (d) $d^2 + 2d + 1$

# 11.6 Brackets

In expressions, calculations inside brackets are always carried out first. The expression $a(b + c)$ means $a \times (b + c)$.

So, if you knew the value of $x$ and had to evaluate the expression $6(x + 1)$, you would first evaluate the bracket and then multiply that value by 6. So, if $x = 3$,

$$6(x + 1) = 6(3 + 1) = 6 \times 4 = 24.$$

---

**Example 11.6.1**

If $a = 8$ and $b = -5$, evaluate    (a) $(2b)^2$,    (b) $(3a + 4b)^2$.

(a) $(2b)^2 = (2 \times (-5))^2 = (-10)^2 = 100$.

(b) $(3a + 4b)^2 = (3 \times 8 + 4 \times (-5))^2 = (24 - 20)^2 = 4^2 = 16$.

---

The approach of Example 11.6.1 is fine if you know the values of the letters, but, if you don't, you need to be able to write expressions such as $4(3y + 2)$ and $4(3y - 2)$ in an alternative form.

Suppose you needed to calculate $2(6 + 3)$; you can see that the answer is $2 \times 9 = 18$. But this is the same as $2 \times 6 + 2 \times 3 = 12 + 6 = 18$. This suggests that $2(6 + 3) = 2 \times 6 + 2 \times 3$, or, in general,

$$a(b + c) = ab + ac.$$

A similar argument would convince you that

$$a(b - c) = ab - ac.$$

If you look at this expression, you can see that each term inside the brackets is multiplied by the term outside.

So $4(3y + 2) = 4 \times (3y) + 4 \times 2 = 12y + 8$, and $4(3y - 2) = 12y - 8$. This process is called **expanding** the brackets.

Suppose now that the sign outside the bracket is negative, and you want to calculate $24 - 2(6 + 3)$. This is $24 - 2 \times 9 = 24 - 18 = 6$. But $24 - 2 \times 6 - 2 \times 3 = 24 - 12 - 6 = 6$, so this suggests that

$$-a(b + c) = -ab - ac.$$

Finally, $24 - 2(6 - 3)$, is equal to $24 - 2 \times 3 = 18$. But you find that $24 - 2 \times 6 + 2 \times 3 = 24 - 12 + 6 = 18$. This suggests that

$$-a(b - c) = -ab + ac.$$

The − sign outside the brackets changes the sign of every term inside the brackets.

---

**Example 11.6.2**

Expand the brackets in these expressions.

(a) $5(x - 2)$  (b) $2(4x - 3)$  (c) $4x(2x - 3)$  (d) $-3(5x - 2)$

(a) $5(x - 2) = 5x - 10$.  (b) $2(4x - 3) = 8x - 6$.

(c) $4x(2x - 3) = 8x^2 - 12x$.  (d) $-3(5x - 2) = -15x + 6$.

---

**Example 11.6.3**

Expand and simplify these expressions.

(a) $3(7x + 4) - 2(5x - 6)$  (b) $2(4x + 5) + (3x - 7)$

(a) $3(7x + 4) - 2(5x - 6) = 21x + 12 - 10x + 12$
$= 11x + 24$.

(b) $2(4x + 5) + (3x - 7)$ means $2(4x + 5) + 1(3x - 7)$. So
$2(4x + 5) + (3x - 7) = 8x + 10 + 3x - 7 = 11x + 3$.

**EXERCISE 11.5**

1 Expand the brackets in these expressions.
  (a) $6(x+3)$       (b) $4(x-1)$       (c) $5(2x+3)$
  (d) $2(3x-5)$     (e) $-4(x+3)$    (f) $-7(x-2)$
  (g) $-8(5x+4)$   (h) $-6(3x-2)$   (i) $x(x+2)$
  (j) $x(4x-3)$     (k) $3x(2x-1)$    (l) $-5x(3x+4)$

2 Expand the brackets in these expressions and simplify the results.
  (a) $3(x-7)+6x$          (b) $8-2(5x-1)$
  (c) $4(x+2)+5(x-3)$    (d) $3(x+4)-2(x-5)$
  (e) $2(6x-5)+3(2x-1)$  (f) $5(4x+3)-3(7x-2)$
  (g) $3(2x+5)+5(4x-3)$  (h) $4(2x-1)-2(4x-5)$
  (i) $x(x+2)-6(x+4)$    (j) $x(x+3)-3(x-8)$
  (k) $x(x-7)+4(2x-3)$   (l) $3x(2x-5)+7(3x+4)$
  (m) $3(6x+2)+(5x-4)$   (n) $8(3x+1)-(6x+7)$
  (o) $5(2x-3)-(10x-9)$  (p) $3+(5x-2)$
  (q) $6x-(6x+5)$        (r) $2(4x+3)-(7x+6)$

## 11.7 Factorizing expressions

**Factorizing** is the opposite process to expanding.

To factorize $7x-21$, the first step is to find a term which is a factor of both $7x$ and $21$ (7 in this case) and this term goes outside the brackets. Then complete the brackets with an expression which, when multiplied by 7, gives $7x-21$. That is, $7x-21=7(x-3)$.

---

**Example 11.7.1**

Factorize  (a) $5x+20$, (b) $6x-9$, (c) $bx-7b$, (d) $x^2+8x$.
  (a) $5x+20=5(x+4)$.    (b) $6x-9=3(2x-3)$.
  (c) $bx-7b=b(x-7)$.    (d) $x^2+8x=x(x+8)$.

---

In these examples, each pair of terms has only one common factor. Sometimes the terms in the expression you want to factorize will have more than one common factor.

In $6x^2+15x$, for example, 3, $x$ and $3x$ are all factors of both $6x^2$ and $15x$. It might seem, therefore, that, if asked to factorize

$6x^2 + 15x$, you could give three possible answers:

$$6x^2 + 15x = 3(2x^2 + 5x),$$

$$6x^2 + 15x = x(6x + 15),$$

$$6x^2 + 15x = 3x(2x + 5).$$

In each of the first two cases, the terms inside the brackets have a common factor but, in the last case, they have not and $6x^2 + 15x$ is said to have been **completely factorized**.

---

**Example 11.7.2**

Factorize completely  (a) $8x^2 - 12x$,  (b) $20ab + 15bc$.

(a) $8x^2 - 12x = 4x(2x - 3)$.  (b) $20ab + 15bc = 5b(4a + 3c)$.

---

### EXERCISE 11.6

**1** Factorize these expressions completely.

| | | |
|---|---|---|
| (a) $3x + 9$ | (b) $14x - 21$ | (c) $20x + 15$ |
| (d) $6x - 15y$ | (e) $ax - 7a$ | (f) $x^2 + 4x$ |
| (g) $12x - 18$ | (h) $15x - 12$ | (i) $4ax + 6bx$ |
| (j) $9x^2 + 12x$ | (k) $20x - 8x^2$ | (l) $4bx - 2bx^2$ |
| (m) $x^2y + xy^2$ | (n) $6ax^2 - 15a^2x$ | (o) $8xy + 12xy^2$ |

## 11.8 Indices

In Section 11.5, you saw how $x \times x$ is written as $x^2$. In a similar way, $x \times x \times x$ is written as $x^3$, which could be read as '$x$ to the power 3' but is usually read as '$x$ cubed'. The 3 is called the **index** or **power**.

$x \times x \times x \times x \times x$ is written as $x^4$, which is read as '$x$ to the power 4'.

Higher powers are written as $x^5$, $x^6$ and so on. $4x^3$ is algebraic shorthand for $4 \times x \times x \times x$. Only the $x$ is cubed, not the 4. In other words, you cube *before* you multiply.

---

**Example 11.8.1**

Write more simply

(a) $c \times c \times c \times c \times c \times c$,  (b) $6 \times d \times d \times d \times d \times d$.

(a) $c \times c \times c \times c \times c \times c = c^6$,  (b) $6 \times d \times d \times d \times d \times d = 6d^5$.

---

### Example 11.8.2

If $p = 6$, evaluate (a) $p^3$, (b) $5p^4$

(a) $p^3 = 6 \times 6 \times 6 = 216$.

(b) $5p^4 = 5 \times 6 \times 6 \times 6 \times 6 = 5 \times 1296 = 6480$.

You can find powers of numbers using the $y^x$ key on your calculator. (Some calculators have differently labelled keys for finding powers of numbers. You may need to use your calculator handbook.)

### Example 11.8.3

If $q = -5$, evaluate $4q^3$.

$4q^3 = 4 \times (-5) \times (-5) \times (-5) = 4 \times (-125) = -500$.

You can use the $\pm$ key on your calculator when finding powers of negative numbers.

### EXERCISE 11.7

1 Write the following in a simpler form.
   (a) $m \times m \times m$        (b) $n \times n \times n \times n \times n$
   (c) $5 \times p \times p \times p \times p \times p$      (d) $3 \times q \times q \times q \times q \times q \times q$

2 If $p = 3$, evaluate (a) $p^5$, (b) $5p^3$.

3 If $q = 10$, evaluate (a) $q^6$, (b) $7q^4$.

4 If $u = -4$, evaluate (a) $u^3$, (b) $3u^5$.

5 If $v = -5$, evaluate (a) $4v^3$, (b) $6v^4$.

## 11.9 Laws of indices

To multiply $a^3$ by $a^4$, you could begin by writing down what each expression means, that is, $a^3 = a \times a \times a$ and $a^4 = a \times a \times a \times a$, so

$$a^3 \times a^4 = a \times a \times a \times a \times a \times a \times a = a^7.$$

The index 7 in the answer is the **sum** $3 + 4$ of the indices 3 and 4 in the question. It suggests the general result

$$a^m \times a^n = a^{m+n}.$$

This result enables you to write down products of powers of the same letter without intermediate working.

**Example 11.9.1**

Simplify $a^2 \times a^6$.

Adding the indices, $a^2 \times a^6 = a^{2+6} = a^8$.

You cannot add the indices unless the letters are the same. You cannot, for example, simplify expressions like $a^3 \times b^4$ by adding the indices.

To divide $a^6$ by $a^2$, note that $a^6 = a \times a \times a \times a \times a \times a$ and $a^2 = a \times a$, so $a^6 \div a^2 = \dfrac{a \times a \times a \times a \times a \times a}{a \times a}$.

You can now 'cancel' the 'fraction' in a way similar to that in which numerical fractions were simplified in Chapter 03, Section 3.4.

$\dfrac{a \times a \times a \times a \times \cancel{a} \times \cancel{a}}{\cancel{a} \times \cancel{a}} = a^4$, that is, $a^6 \div a^2 = a^{6-2} = a^4$.

The index 4 in the answer is the difference $6 - 2$ between the indices in the question. It suggests the general result

$a^m \div a^n = a^{m-n}$, provided that $m$ is greater than $n$.

Similarly, if you were dividing $a^2$ by $a^6$, you would find that

$a^2 \div a^6 = \dfrac{a \times a}{a \times a \times a \times a \times a \times a} = \dfrac{1}{a^4}$.

This suggests the general result

$a^m \div a^n = \dfrac{1}{a^{n-m}}$, provided that $n$ is greater than $m$.

Using these general results, you can divide powers of the same letter.

**Example 11.9.2**

Simplify (a) $a^9 \div a^4$, (b) $a^3 \div a^8$

(a) Subtracting the indices, $a^9 \div a^4 = a^5$.

(b) $a^3 \div a^8 = \dfrac{1}{a^{8-3}} = \dfrac{1}{a^5}$.

Using the general result to simplify $a^4 \div a^3$ gives the answer $a^1$ but

$$a^4 \div a^3 = \frac{a \times a \times a \times a}{a \times a \times a} = a \text{ so } a^1 = a.$$

### Example 11.9.3

Simplify (a) $a \times a^3$, (b) $a^3 \div a$.

(a) $a \times a^3 = a^1 \times a^3 = a^4$. (b) $a^3 \div a = a^3 \div a^1 = a^2$.

To simplify $(a^4)^3$, write it as $(a^4)^3 = a^4 \times a^4 \times a^4 = a^{4 \times 3} = a^{12}$.

This suggests the general result $(a^m)^n = a^{mn}$.

### Example 11.9.4

Simplify $(a^3)^5$.

Multiplying the indices, $(a^3)^5 = a^{3 \times 5} = a^{15}$.

## Summary

$a^m \times a^n = a^{m+n}$

$a^m \div a^n = a^{m-n}$, if $m$ is bigger than $n$.

$a^m \div a^n = \dfrac{1}{a^{n-m}}$ if $n$ is bigger than $m$.

$(a^m)^n = a^{mn}$.

### EXERCISE 11.8

1 Simplify the following where possible.

(a) $x^3 \times x^4$ (b) $x^7 \div x^2$ (c) $(x^4)^3$ (d) $x \times x^5$
(e) $x^9 \div x^4$ (f) $(x^5)^2$ (g) $x^4 \times x^6$ (h) $x^3 \times y^4$
(i) $(x^2)^2$ (j) $x^6 \div x$ (k) $x^5 \div x^3$ (l) $x^3 \times x$
(m) $x^6 \div x^5$ (n) $x^7 \times x^3$ (o) $x^3 \div x$ (p) $(x^3)^3$

## 11.10 Simplifying expressions with indices

To simplify $4a^3 \times 5a^2$, consider first what it means.

$$4a^3 \times 5a^2 = (4 \times a^3) \times (5 \times a^2) = (4 \times 5) \times (a^3 \times a^2) = 20a^5.$$

This result shows that you can consider the numbers and the powers of $a$ separately. Since $4 \times 5 = 20$ and $a^3 \times a^2 = a^5$, $4a^3 \times 5a^2 = 20a^5$. You can use this method to simplify expressions with more letters.

**Example 11.10.1**

Simplify (a) $3a^2b^4 \times 6a^5b^2$, (b) $3a^5 \times 4a$, (c) $7a^3 \times a^2$, (d) $8a^5 \times a$.

(a) $3a^2b^4 \times 6a^5b^2 = (3 \times 6) \times (a^2 \times a^5) \times (b^4 \times b^2) = 18a^7b^6$.
(b) $3a^5 \times 4a = 12a^6$. (Remember $4a$ means $4a^1$.)
(c) $7a^3 \times a^2 = 7a^5$.
(d) $8a^5 \times a = 8a^6$.

To simplify $12a^5 \div 3a^2$, you can again consider the numbers and the powers of $a$ independently.

$$12a^5 \div 3a^2 = \frac{12 \times a^5}{3 \times a^2} = \frac{12}{3} \times \frac{a^5}{a^2} = 4a^3.$$

**Example 11.10.2**

Simplify (a) $24a^6 \div 8a^4$, (b) $\dfrac{18a^8}{3a^5}$

(a) $24a^6 \div 8a^4 = 3a^2$. (b) $\dfrac{18a^8}{3a^5} = 6a^3$.

To simplify $\dfrac{10a^4b^5}{2a^3b^2}$,

$$\frac{10a^4b^5}{2a^3b^2} = \frac{10}{2} \times \frac{a^4}{a^3} \times \frac{b^5}{b^2} = 5ab^3.$$

**Example 11.10.3**

Simplify (a) $\dfrac{8a^5b^3}{2a^3b}$, (b) $5a^6b^7 \div a^4b^3$, (c) $\dfrac{15a^5b^2}{3a^2c^3}$.

(a) $\dfrac{8a^5b^3}{2a^3b}, = \dfrac{8}{2} \times \dfrac{a^5}{a^3} \times \dfrac{b^3}{b} = 4a^2b^2$.

(b) $5a^6b^7 \div a^4b^3 = \dfrac{5a^6b^7}{a^4b^3} = \dfrac{5}{1} \times \dfrac{a^6}{a^4} \times \dfrac{b^7}{b^3} = 5a^2b^4$.

(c) $\dfrac{15a^5b^2}{3a^2c^3} = \dfrac{15}{3} \times \dfrac{a^5}{a^2} \times \dfrac{b^2}{1} \times \dfrac{1}{c^3} = \dfrac{5a^3b^2}{c^3}$.

You may find it helpful at first to write the expressions out in full.

---

**Example 11.10.4**

Simplify (a) $\dfrac{4a^2}{2a^5}$, (b) $\dfrac{6a^8b^3}{2a^5b^7}$, (c) $\dfrac{3a^5b^2}{6a^4b^4}$, (d) $\dfrac{14a^3b^2}{2b^2c}$.

(a) $\dfrac{4a^2}{2a^5} = \dfrac{4}{2} \times \dfrac{a^2}{a^5} = \dfrac{2}{a^3}$.

(b) $\dfrac{6a^8b^3}{2a^5b^7} = \dfrac{6}{2} \times \dfrac{a^8}{a^5} \times \dfrac{b^3}{b^7} = \dfrac{3a^3}{b^4}$.

(c) $\dfrac{3a^5b^2}{6a^4b^4} = \dfrac{3}{6} \times \dfrac{a^5}{a^4} \times \dfrac{b_2}{b^4} = \dfrac{a}{2b^2}$.

(d) $\dfrac{14a^3b^2}{2b^2c} = \dfrac{14}{2} \times \dfrac{a^3}{1} \times \dfrac{b^2}{b^2} \times \dfrac{1}{c} = \dfrac{7a^3}{c}$, as $\dfrac{b^2}{b^2} = 1$.

---

To simplify $(3a^4)^2$, a good method is to write it as $3a^4 \times 3a^4$, which is $9a^8$, that is, both the 3 and the $a^4$ have been squared. With practice, you should be able to write the answer straight down.

With higher powers, it can be easier to use the index law $(a^m)^n = a^{mn}$. Thus, $(5a^2b)^3 = 5^3 \times (a^2)^3 \times b^3 = 125a^6b^3$. Note that each term inside the bracket, that is, 5, $a^2$ and $b$, has been cubed.

---

**Example 11.10.5**

Simplify (a) $(5a)^2$, (b) $(4a^2b^3)^2$, (c) $(2a^4)^3$, (d) $(4a^2b^5)^3$.

(a) $(5a)^2 = 5a \times 5a = 25a^2$.
(b) $(4a^2b^3)^2 = 4a^2b^3 \times 4a^2b^3 = 16a^4b^6$.
(c) $(2a^4)^3 = 2^3 \times (a^4)^3 = 8a^{12}$.
(d) $(4a^2b^5)^3 = 4^3 \times (a^2)^3 \times (b^5)^3 = 64a^6b^{15}$.

---

## EXERCISE 11.9

**1** Simplify the following.

(a) $3x^2 \times 2x^3$  (b) $5x \times 4x^2$  (c) $7x^2y^4 \times 2x^3y^2$

(d) $9x^7 \times x$  (e) $6x^4 \times x^5$  (f) $8x^3y^3 \times x^4y^2$

(g) $12x^5 \div 4x^3$  (h) $\dfrac{21x^7}{3x^4}$  (i) $\dfrac{10x^6y^4}{2x^4y^2}$

2 Simplify the following.

(a) $\dfrac{24x^5y^4}{8x^4y}$    (b) $6x^7y^6 \div x^2y^4$    (c) $\dfrac{10x^7y^3}{2x^3z^2}$

(d) $\dfrac{a^4}{a^9}$    (e) $\dfrac{16x^2}{2x^6}$    (f) $\dfrac{8x^6y^3}{2x^2y^5}$

(g) $\dfrac{5x^3y^6}{15x^4y^2}$    (h) $\dfrac{20x^3y^4}{4x^3}$    (i) $(6x)^2$

(j) $(5xy)^2$    (k) $(x^5y^3)^2$    (l) $(5x^3y^4)^2$

(m) $(10x)^3$    (n) $(4x^5)^3$    (o) $(5x^4y^3)^3$

# 12

# approximation

**In this chapter you will learn:**

- how to round whole numbers and decimals
- about significant figures
- how to find estimates
- how to use rounding in problems
- about the accuracy of measurements.

# 12.1 Introduction

Sometimes exact answers to a problem are unnecessary, or even impossible, and a sensible *approximate* answer is needed. For example, in a census, the population of Coventry was given as 304 426 but, for most practical purposes, 300 000 would be an adequate approximation.

You will meet a variety of ways of approximating numbers. Then you will use approximate values to find estimates to calculations, to solve problems and see how approximation and measurement are related.

# 12.2 Rounding whole numbers

One method of approximating is to give, or **round**, the number to the nearest 10, 100, 1000 etc.

For example, 22 is between 20 and 30 but it is nearer to 20 than 30. So, when you round 22 to the nearest 10, you **round down** to get 20. When you round 67 to the nearest 10, however, you **round up** to get 70, as 67 is nearer to 70 than to 60.

If a number is exactly halfway between two multiples of 10, the convention is to **round up**. For example, 85 is exactly halfway between 80 and 90 so, to round 85 to the nearest 10, round up to 90.

In other words, if the digit in the units column is less than 5, you do not alter any of the previous digits but, if the digit in the units column is 5 or more, you increase the tens digit by one. In both cases, your answer will have a zero in the units column.

---

**Example 12.2.1**

Round each of the following to the nearest ten:
(a) 392,   (b) 4697.

(a) 392 is between 390 and 400. The 2 tells you to round down, so 392 to the nearest ten is 390.
(b) 4698 is between 4690 and 4700. The 8 tells you to round up, so 4698 to the nearest ten is 4700.

---

Similarly, you can round numbers to the nearest hundred, when the tens digit tells you whether to round up or down, or to the nearest thousand, when it is the hundreds digit which tells you.

**Example 12.2.2**

Round to the nearest hundred (a) 349, (b) 2861, (c) 31 750.

(a) 349 is between 300 and 400. The 4 tells you to round down, so 349 to the nearest hundred is 300.
(b) 2861 is between 2800 and 2900. The 6 tells you to round up, so 2861 to the nearest hundred is 2900.
(c) 31 750 is between 31 700 and 31 800. The 5 tells you to round up, so 31 750 to the nearest hundred is 31 800.

## EXERCISE 12.1

1 Write each of these numbers to the nearest ten.
(a) 34      (b) 45      (c) 76
(d) 751     (e) 8465    (f) 7396

2 Write each of these numbers to the nearest hundred.
(a) 675     (b) 849     (c) 350
(d) 4351    (e) 2974    (f) 19 483

3 Write each of these numbers to the nearest thousand.
(a) 7682    (b) 3429    (c) 5500
(d) 37 743  (e) 42 499  (f) 99 682

4 The attendance at a Manchester United soccer match was 61 267. Round this number to the nearest hundred.

5 In a census, the population of Wolverhampton was given as 242 187. Round this number
(a) to the nearest thousand,      (b) to the nearest hundred thousand.

6 To the nearest hundred, the population of a town is 7600. Find
(a) the smallest,      (b) the largest,
possible population the town could have.

## 12.3 Rounding with decimals

If you travel 70 miles in $1\frac{1}{2}$ hours, you can find your average speed (see page 91) by dividing 70 by 1.5. My calculator gives 46.66666667 as the answer to $70 \div 1.5$ but it would not be sensible to write down the whole calculator display. An answer to the nearest **whole number** is appropriate, so as 46.66666667 is between 46 and 47, but nearer to 47, an average speed of 47 mph is a reasonable answer.

If the digit in the first decimal place is less than 5, you round down, that is, you leave the whole number unchanged but, if it is 5 or more, you round up, that is, you increase the whole number by one.

---

**Example 12.3.1**

Write each of the following to the nearest whole number.
(a) 4.369   (b) 64.7132   (c) 368.5

(a) 4.369 is between 4 and 5. The 3 tells you to round down, so 4.369 to the nearest whole number is 4.
(b) 64.7132 is between 64 and 65. The 7 tells you to round up, so 64.7132 to the nearest whole number is 65.
(c) 368.5 is halfway between 368 and 369. The 5 tells you to round up, so 368.5 to the nearest whole number is 369.

---

For greater accuracy, you can give numbers **correct to one decimal place** (1 d.p.). 7.346, for example, is between 7.3 and 7.4, but nearer to 7.3, so 7.346 = 7.3 (to 1 d.p.).

If the digit in the *second* decimal place (4 in 7.346) is less than 5, round down, leaving all the preceding digits unchanged but, if it is 5 or more, round up, increasing the digit in the first decimal place by one.

---

**Example 12.3.2**

Write each of these numbers correct to one decimal place.
(a) 7.483   (b) 0.65   (c) 23.0481.

(a) 7.483 is between 7.4 and 7.5; the 8 tells you to round up, so 7.483 = 7.5 (to 1 d.p.)
(b) 0.65 is halfway between 0.6 and 0.7; the 5 tells you to round up, so 0.65 = 0.7 (to 1 d.p.)
(c) The 4 in 23.0481 tells you that 23.0481 = 23.0 (to 1 d.p.).

---

In the answer 23.0, the zero *is* necessary; 23 alone is correct only to the nearest whole number.

When you express a number correct to one decimal place, you are giving it correct to the nearest *tenth*.

In a similar way, you can approximate to any number of decimal places. To give a number correct to two decimal places (nearest hundredth), look at the digit in the *third* decimal place; to give a number correct to three decimal places (nearest thousandth), look at the digit in the *fourth* decimal place and so on.

---

**Example 12.3.3**

(a) Write 24.6472 correct to two decimal places.
(b) Write 0.063 279 correct to three decimal places.

(a) The digit in the *third* decimal place is 7 so, rounding up, 24.6472 = 24.65 (to 2 d.p.).
(b) The digit in the *fourth* decimal place is 2 so, rounding down, 0.063 279 = 0.063 (to 3 d.p.).

---

**Example 12.3.4**

Write 3.704891 correct to (a) to 3 d.p. (b) to 2 d.p.

(a) 3.704891 = 3.705 (to 3 d.p.).
(b) 3.704891 = 3.70 (to 2 d.p.).

---

Notice that, if you give the answer to (a) correct to two decimal places, you get 3.71, which is *not* the correct answer to (b).

### EXERCISE 12.2

1 Write each of the following to the nearest whole number.
   (a) 8.74         (b) 21.241      (c) 67.5
   (d) 260.2841     (e) 399.72

2 Write each of these numbers correct to one decimal place.
   (a) 9.927        (b) 0.8643      (c) 17.9821
   (d) 76.85        (e) 400.0389

3 Write each of these numbers correct to two decimal places.
   (a) 4.2491       (b) 8.465       (c) 0.748 36
   (d) 0.061 794

4 Write each of these numbers correct to three decimal places.
   (a) 8.678 23     (b) 7.608 52    (c) 13.419 61
   (e) 0.009 472

5  Calculate the value of $86 \div 9$ and give your answer to the nearest whole number.

6  Calculate the value of $8.3 \div 3.9$ and give your answer correct to the nearest tenth.

7  Calculate the value of $4.68 \times 3.72$ and give your answer correct to the nearest hundredth.

8  Convert $\frac{3}{7}$ to a decimal and give your answer correct to three decimal places.

9  Correct to one decimal place, a number is 7.3. Write down the smallest possible value of the number.

10  A number which is divisible by 5 has three decimal places. Correct to one decimal place, the number is 5.8 and, correct to two decimal places, it is 5.85. Find the number.

# 12.4  Significant figures

Giving a number to a specified number of **significant figures** is another method of approximating. For example, in the number 7483, the most *significant*, or important, figure is 7, as its value is 7000. To give 7483 correct to one significant figure (1 s.f.), you have to choose between 7000 and 8000, the rounded up value.

To decide which it is, use the same rules you have used earlier in this chapter; you look at the *second* figure, 4, which tells you to leave the 7 unchanged. So, $7483 = 7000$ (to 1 s.f.), the same answer as giving it to the nearest thousand. Notice that, when you give a number correct to one significant figure, there is only one non-zero figure in your answer.

The second most significant figure in the number 7483 is 4, as its value is 400. To give 7483 correct to two significant figures (2 s.f.), you have to choose between 7400 and 7500. The third figure, 8, tells you to round up. So $7483 = 7500$ (to 2 s.f.), the same answer as giving it to the nearest hundred.

When you give a number correct to two significant figures, there are usually two non-zero figures in your answer but there could be only one. For example, $698 = 700$ (to 2 s.f.).

For numbers less than 1, zeros at the start of the number are *not* significant. For example, to give 0.0543 correct to one significant figure, the 4 tells you not to round up and so $0.0543 = 0.05$ (to 1 s.f.), the same answer as giving it to two decimal places. As before, there is only one non-zero figure in your answer.

### Example 12.4.1

Write each of these numbers correct to one significant figure. (a) 5.23 (b) 2500 (c) 0.783 (d) 0.000 624

(a) 5.23 = 5 (to 1 s.f.).     (b) 2500 = 3000 (to 1 s.f.).
(c) 0.783 = 0.8 (to 1 s.f.).   (d) 0.000 624 = 0.0006
                                      (to 1 s.f.).

### Example 12.4.2

Write each of these numbers correct to three significant figures.
(a) 23.8486 (b) 0.006 083 4 (c) 0.000 379 62

(a) 23.8486 = 23.8 (to 3 s.f.).
(b) 0.006 083 4 = 0.006 08 (to 3 s.f.).
(c) 0.000 379 62 = 0.000 380 (to 3 s.f.).

### EXERCISE 12.3

1 Write each of these numbers correct to one significant figure.
   (a) 23.6        (b) 375        (c) 0.187
   (d) 0.0072      (e) 0.085

2 Write each of these numbers correct to two significant figures.
   (a) 8673         (b) 742.9        (c) 6.85
   (d) 0.084 72     (e) 0.003 971

3 Write each of these numbers correct to three significant figures.
   (a) 23 679       (b) 376.47       (c) 13.7819
   (d) 8.925        (e) 0.084 382    (f) 0.000 436 5
   (g) 0.090 237    (h) 0.003 697 4

4 Calculate the value of $43.7 \times 36.8$ and give your answer correct to one significant figure.

5 Calculate the value of $41 \div 7$ and give your answer correct to two significant figures.

6 Calculate the value of
   (a) $3.82 \times 2.97$,     (b) $4.72 \div 9.84$.

   Give your answers correct to three significant figures.

7 Convert $\frac{1}{11}$ to a decimal and give your answer correct to three significant figures.

8 Write 599.9 correct to
   (a) 1 s.f.    (b) 2 s.f.    (c) 3 s.f.
9 Write 0.059 99 correct to
   (a) 1 s.f.    (b) 2 s.f.    (c) 3 s.f.

## 12.5 Estimates

You should always make an estimate to check that your answer to a calculation is sensible. To do this, approximate to each of the numbers involved and do the calculation mentally using these approximations.

For example, to estimate the answer to $27.36 \times 0.241$, you could say that 27.36 is near 28 and 0.241 is near $\frac{1}{4}$. The product, therefore, is near $28 \times \frac{1}{4}$, which is 7. The exact answer must be less than 7, because 28 and $\frac{1}{4}$ are both greater than the original numbers.

The usual method of estimating, however, is to give each number correct to one significant figure.
So $27.36 \times 0.241 \approx 30 \times 0.2 = 6$. (Recall that the symbol $\approx$ means 'is approximately equal to'.)

---

**Example 12.5.1**

Find an estimate for    (a) $76.48 \div 2.32$,    (b) $\dfrac{8.67 \times 0.235}{0.058}$

(a) $76.48 \div 2.32 \approx 80 \div 2 = 40$.

(b) $\dfrac{8.67 \times 0.235}{0.058} \approx \dfrac{9 \times 0.2}{0.06} = \dfrac{1.8}{0.06} = \dfrac{180}{6} = 30$.

---

### EXERCISE 12.4

1 Find an estimate for each of these calculations and then use a calculator to find the answer. Where necessary, give your answer correct to three significant figures.

   (a) $2.39 \times 3.74$          (b) $34.91 \times 2.83$
   (c) $31 \div 5.83$             (d) $247.1 \times 47.8$
   (e) $543 \div 97.4$            (f) $76.4 \div 23.2$
   (g) $32.8 \times 3.74 \times 66.8$     (h) $\dfrac{41.7 \times 8.92}{27.4}$

(i) $67.3 \times 0.347$
(k) $56.4 \div 0.286$
(m) $0.0673 \times 0.002\ 36$
(o) $0.0437 \div 0.176$

(j) $0.628 \div 1.84$
(l) $376 \times 0.0354$
(n) $8.32 \div 0.381$
(p) $0.783 \div 0.0154$

(q) $0.0643 \div 2.816$

(r) $\dfrac{4.79 \times 0.0783}{0.247}$

(s) $\dfrac{0.42 \times 0.673}{0.0732}$

(t) $\dfrac{369 \times 0.527}{98.3 \times 0.0162}$

## 12.6 Rounding in practical problems

So far you have used rules to decide whether to round up or down but there are many real life situations in which these rules are not relevant and you have to use common sense.

For example, if you had 375 eggs and wanted to know how many boxes, each holding six eggs, you could fill, you would probably use your calculator to work out $375 \div 6$, obtaining the answer 62.5. To the nearest whole number 62.5 is 63, but there are not enough eggs to fill the sixty third box and so you round down to 62.

The 62 full boxes contain 372 eggs ($62 \times 6$), so there are 3 eggs left.

### EXERCISE 12.5

1 How many boxes, each holding 6 eggs, can be filled with 250 eggs? How many eggs are left over?

2 How many pieces of metal 7 cm long can be cut from bar 60 cm long? What length of metal is left?

3 The emulsion paint in a 1 litre tin covers an area of about $12\ m^2$. How many tins of paint are needed to cover an area of $45\ m^2$?

4 The maximum number of people allowed in a lift is 30. Find the minimum number of trips the lift must make to carry 65 people.

5 A decorator needs 27 litres of masonry paint for the outside of a house. The paint is sold in 5 litre tins. How many tins must he buy?

6 How many cars, each 4.5 m long, would fit end to end on a ferry deck which is 31 m long?

7  190 tiles are needed for a bathroom. The tiles are sold in boxes of 20. How many boxes are needed?

8  A railway carriage has seats for 56 passengers. How many carriages are needed for 400 passengers?

## 12.7  Accuracy of measurements

All measurements are approximate, even if you are very careful when making them and use sophisticated measuring instruments. For example, giving the length of a line as '7 cm correct to the nearest centimetre' means that its length lies between 6.5 cm and 7.5 cm, that is, 6.5 cm ≤ actual length < 7.5 cm. So the maximum possible error is 0.5 cm, which is half the unit being used for the measurement.

The length of the line might be given more accurately to the nearest 0.1 centimetre, or millimetre, as 7.3 cm. This means that its actual length lies between 7.25 cm and 7.35 cm and so the maximum possible error is 0.05 cm, which is again half the unit of measurement.

Even if the degree of accuracy is not explicitly stated, it can be implied by the number of decimal places given in the measurement.

---

**Example 12.7.1**

The world record for the women's 100 metres is 10.49 seconds.
(a)  To what degree of accuracy is this time given?
(b)  Write down the minimum time and the maximum time this could be.
(c)  What is the maximum possible error?

(a)  10.49 has two decimal places and so the time is correct to the nearest 0.01 second.
(b)  The minimum time is 10.485 s; the maximum time is 10.495 s.
(c)  The maximum possible error is 0.005 s.

---

Note that in part (b) you may think, reasonably, that the maximum time should not be 10.495 s because that would correct to 10.50 s, and that it should be the largest number which is just less than 10.495. However, there is no such number. No matter

how close a number to 10.495 you choose, say 10.494 $\overbrace{999\cdots9}$, there is always a closer number to 10.495, obtained by taking one more 9. Thus it is conventional to say that the maximum is 10.495, although strictly this is not true.

## EXERCISE 12.6

1 The distance from London to Liverpool is 325 km, correct to the nearest kilometre. Write down the minimum distance and the maximum distance this could be.

2 The largest butterfly in the United Kingdom has a wingspan of 12.7 cm, correct to the nearest 0.1 cm. Write down the minimum length and the maximum length this could be.

3 The world record for the men's 100 metres is 9.85 seconds, correct to the nearest 0.01 second. Write down the minimum time and the maximum time this could be.

4 The heaviest baby on record weighed 10.6 kg at birth.
   (a) To what degree of accuracy is this weight given?
   (b) Write down the minimum and the maximum weight this could be.
   (c) What is the maximum possible error in recording the weight as 10.6 kg?

5 The world record for the women's 400 metres is 47.60 seconds.
   (a) To what degree of accuracy is this time given?
   (b) What is the maximum possible error?

**equations 1**

**In this chapter you will learn:**

- how to solve linear equations
- how to solve problems using equations
- how to solve inequalities.

# 13.1 Introduction

An **equation** is a mathematical statement that two expressions are equal, such as $3 \times 2 = 6$. Sometimes, missing numbers are denoted by symbols or letters, for example, $\square + 2 = 7$ or $x - 3 = 2$. In both cases, the equation is **satisfied**, or true, if you replace the missing number by 5, which is called the **solution**, or **root**, of the equation.

Mathematicians in Ancient Egypt and Babylon were able to solve many types of equation, which were expressed in words or drawings. In Greece, Diophantus (c 275 BC), who has been called the 'Father of Algebra', used a form of symbolism, although the notation in use today did not appear until the seventeenth century.

In the Islamic world of the Middle Ages, equation solving was known as 'the science of restoration and balancing'. The Arabic word for restoration, *al-jabru*, is the root of 'algebra'. Even up to the sixteenth century, algebra was the name given to the art of bonesetting.

# 13.2 Finding missing numbers

'$\square + 7 = 16$. What number makes this equation true?'

'I think of a number and add 7 to it. The result is 16. Find the number.'

'$x + 7 = 16$. What number does $x$ stand for?'

These are three different ways of asking the same question. You can spot by mental arithmetic that the unknown number in each case is 9.

### EXERCISE 13.1

1 Find the numbers which make these equations true.
   (a) $\square + 8 = 13$   (b) $\square - 5 = 6$   (c) $\square \times 4 = 36$
   (d) $\square \div 3 = 4$   (e) $10 - \square = 3$   (f) $(2 \times \square) + 3 = 11$
   (g) $3 \times (\square + 1) = 9$   (h) $(\square - 3) \times 5 = 5$   (i) $(8 + \square) \div 5 = 3$

2 Find the number in each of the following.
   (a) I think of a number and add 9 to it. The result is 22.
   (b) I think of a number and subtract 5 from it. The result is 12.
   (c) I think of a number and multiply it by 6. The result is 42.
   (d) I think of a number and divide it by 3. The result is 9.
   (e) I think of a number, double it and add 3. The result is 15.

3 Find the number which $x$ stands for in each of these equations.
   (a) $x + 3 = 7$   (b) $x - 2 = 4$   (c) $5x = 20$   (d) $\frac{1}{4}x = 3$

# 13.3 Solving linear equations

In this context, linear means that the equation does not involve higher powers of $x$, such as $x^2$ or $x^3$.

If you had to solve the equation $x + 4 = 9$, in other words, find the number which $x$ stands for, you would probably just say '$x = 5$, because $5 + 4 = 9$.' With more complicated equations, however, you will not be able to write answers straight down; more systematic methods are needed.

In this case, you can subtract 4 from both sides of the equation $x + 4 = 9$. This gives $x = 9 - 4$, which leads to the correct answer.

You may find it helpful to think of an equation as a balance, with $x + 4$ in one scale pan and 9 in the other. If you subtract 4 from both sides of the equation $x + 4 = 9$ you get $x = 9 - 4$, giving $x = 5$. The idea of a balance is key, and may be helpful in deciding how to tackle a problem.

---

**Example 13.3.1**

Solve the equation $x - 5 = 3$.
Add 5 to each side $\qquad x = 3 + 5$
$\qquad\qquad\qquad\qquad\quad x = 8.$

---

You should always check your solution by substituting it back into the original equation. In this case, $8 - 5$ does equal 3.

The solution of the equation $3x = 15$ is $x = 5$. To obtain this, you could divide both sides of the equation by 3. This gives $x = \frac{15}{3}$, which is equal to 5. In terms of the balance, if two masses, $3x$ and 15, are equal, then $\frac{1}{3}$ of each of them will also be equal. Checking the solution, $3 \times 5 = 15$.

---

**Example 13.3.2**

Solve the equation $4x = 11$ and check your answer.
Divide both sides by 4 $\qquad x = \frac{11}{4}$
$\qquad\qquad\qquad\qquad\qquad\quad x = 2\frac{3}{4}.$

Check $4 \times 2\frac{3}{4} = 11.$

Multiplication and division are the inverse operations of each other.

---

### Example 13.3.3

Solve the equation $\frac{1}{6}x = 4$ and check your answer.

Multiply both sides by 6 $\quad x = 4 \times 6$

$$x = 24.$$

Check $\frac{1}{6} \times 24 = 4.$

---

In general, you must do the same to both sides of an equation so that the balance of the equation is not destroyed.

### EXERCISE 13.2

1 Solve these equations and check your answers.

(a) $x + 4 = 7$ (b) $x - 3 = 5$ (c) $4x = 24$ (d) $\frac{1}{5}x = 2$

(e) $x + 6 = 1$ (f) $5x = 3$ (g) $x + 4 = 4$ (h) $3x = 7$

(i) $7x = 0$ (j) $3 + x = -2$ (k) $x - 3 = -1$ (l) $\frac{1}{6}x = -1$

With a more complicated equation, you should aim to rearrange it so that the $x$ term is on its own on one side of the equation and there is a number on the other side. For example, to solve $4x + 9 = 1$, *subtract* 9 from each side, that is, $4x = 1 - 9$ or $4x = -8$. Then *divide* both sides by 4 to obtain the solution $x = -\frac{8}{4}$, or $x = -2$.

In the equation $4x + 9 = 1$, the number, 4, multiplying $x$ is called the **coefficient** of $x$.

---

### Example 13.3.4

Solve the equation $6x - 1 = 3$ and check your answer.

Add 1 to each side $\quad 6x = 3 + 1$

$$6x = 4$$

Divide both sides by 6 $\quad x = \frac{4}{6} = \frac{2}{3}$

Check $\left(6 \times \frac{2}{3}\right) - 1 = 4 - 1 = 3.$

---

Sometimes an equation may have terms involving $x$ on both sides. To solve such an equation, start by collecting all the $x$ terms on one side and then carry on in the usual way.

### Example 13.3.5

Solve the equation $5x + 1 = 2x - 2$ and check your answer.

$$5x + 1 = 2x - 2$$

Subtract $2x$ from both sides $\qquad 5x - 2x + 1 = -2$

$$3x + 1 = -2$$

Subtract 1 from both sides $\qquad\qquad 3x = -2 - 1$

$$3x = -3$$

Divide both sides by 3 $\qquad\qquad\qquad x = -1.$

Check LHS (left-hand side) $= (5 \times (-1)) + 1 = -5 + 1 = -4.$

RHS (right-hand side) $= (2 \times (-1)) - 2 = -2 - 2 = -4.$

Note that you could have started by subtracting $5x$ from both sides of the equation. In that case you would have reached the equation $-3x = 3$. Dividing both sides by $-3$ gives $x = -1$, as before.

To solve an equation with a negative $x$ term, make it zero by *adding* an appropriate $x$ term to both sides. To solve $10 - 3x = 5$, for example, start by adding $3x$ to both sides.

### Example 13.3.6

Solve $10 - 3x = 5$.

Add $3x$ to both sides $\qquad\qquad 10 = 3x + 5$

Subtract 5 from both sides $\qquad 5 = 3x$

Divide both sides by 3 $\qquad\qquad x = \frac{5}{3} = 1\frac{2}{3}.$

Check $10 - (3 \times 1\frac{2}{3}) = 10 - 5 = 5.$

Alternatively, you could start by subtracting 10 from both sides giving $-3x = -5$ and then divide both sides by $-3$. This is a correct method for solving equations but it has the disadvantage that arithmetical errors are more likely with negative numbers.

For an equation with $x$ terms on both sides but just one negative coefficient, you can use the same approach as in Example 13.3.6.

### Example 13.3.7

Solve $1 - 2x = 3x + 8$ and check your answer.

Add $2x$ to both sides $\qquad\qquad 1 = 5x + 8$

Subtract 8 from both sides $\qquad -7 = 5x$

Divide both sides by 5 $\qquad x = -\frac{7}{5} = -1\frac{2}{5}.$

Check LHS $= 1 - \left(2 \times \left(-1\frac{2}{5}\right)\right) = 1 + 2\frac{4}{5} = 3\frac{4}{5}.$

RHS $= \left(3 \times \left(-1\frac{2}{5}\right)\right) + 8 = -4\frac{1}{5} + 8 = 3\frac{4}{5}.$

To solve equations with more than one negative coefficient, add to each side an $x$ term which is big enough to give you an equation with no negative coefficients. For example, to solve $8 - 3x = 2 - 5x$, you start by adding $5x$ to both sides. If you added only $3x$, the equation you obtain still has a negative coefficient.

**Example 13.3.8**

Solve $8 - 3x = 2 - 5x$.

Add $5x$ to both sides $\qquad 8 + 2x = 2$
Subtract 8 from both sides $\quad 2x = -6$
Divide both sides by 5 $\qquad x = -3.$
Check LHS $= 8 - (3 \times (-3)) = 8 - (-9) = 8 + 9 = 17.$
RHS $= 2 - (5 \times (-3)) = 2 - (-15) = 2 + 15 = 17.$

## EXERCISE 13.3

1 Solve these equations and check your answers.
   (a) $2x + 1 = 9$ (b) $3x - 2 = 7$ (c) $4x + 3 = 5$
   (d) $5x + 8 = 3$ (e) $6x - 5 = 9$ (f) $8x + 9 = 3$
   (g) $5 + 3x = 17$ (h) $7x + 9 = 9$ (i) $2x + 3 = 8$
   (j) $6x - 1 = 4$ (k) $4x - 1 = -9$ (l) $3x + 7 = -1$

2 Solve these equations and check your answers.
   (a) $3x + 1 = 2x + 5$ (b) $5x - 3 = 3x + 7$
   (c) $6x - 2 = 4x - 1$ (d) $7x + 3 = 3x - 5$
   (e) $5x - 2 = 8x - 7$ (f) $2x + 1 = 7x + 3$
   (g) $9x + 4 = 5x - 6$ (h) $8x + 1 = 2x + 5$
   (i) $7x + 1 = 5x + 1$ (j) $5x + 6 = 2x - 3$
   (k) $7x + 1 = x - 3$ (l) $10x - 9 = 2x + 13$

3 Solve these equations and check your answers.
   (a) $8 - x = 3$ (b) $7 - x = 12$
   (c) $8 - 3x = 2$ (d) $9 - 5x = 6$
   (e) $1 - 4x = 9$ (f) $8 - 3x = 1$

(g) $3 - 2x = 4$      (h) $3 - 4x = 9$
(i) $x + 4 = 7 - 2x$      (j) $2x + 3 = 7 - 3x$
(k) $3 - 4x = 5x + 12$      (l) $4 - 7x = 3x + 4$
(m) $8 - 3x = 2x - 5$      (n) $7x + 6 = 2 - 3x$
(o) $4 - x = 7 - 2x$      (p) $9 - 2x = 1 - 4x$

## 13.4 Equations with brackets

To solve an equation which has brackets, the usual first step is to expand the brackets. Sometimes, this step alone transforms the equation into one of the types you have already met.

---

**Example 13.4.1**

Solve $4(2x - 3) = 20$.

$$4(2x - 3) = 20$$

Expand the brackets      $8x - 12 = 20$
Add 12 to both sides      $8x = 32$
Divide both sides by 8      $x = 4$

Check $4(2 \times 4 - 3) = 4(8 - 3) = 4 \times 5 = 20$.

---

If necessary, after expanding the brackets, simplify both sides of the equation by collecting like terms.

---

**Example 13.4.2**

Solve $7 + 2(3x - 1) = 8x - 3(2x + 5)$.

$$7 + 2(3x - 1) = 8x - 3(2x + 5)$$

Expand the brackets      $7 + 6x - 2 = 8x - 6x - 15$
Collect like terms      $5 + 6x = 2x - 15$
Subtract $2x$ from both sides      $5 + 4x = -15$
Subtract 5 from both sides      $4x = -20$
Divide both sides by 4      $x = -5$.
Check LHS $= 7 + 2(3 \times (-5) - 1) = 7 + 2 \times (-16) = -25$.
     RHS $= 8 \times (-5) - 3(2 \times (-5) + 5) =$
     $-40 - 3 \times (-5) = -25$.

---

## EXERCISE 13.4

1 Solve these equations and check your answers.
     (a) $5(x + 4) = 35$      (b) $3(4x - 5) = 9$
     (c) $4(x + 6) = 12$      (d) $3(2x - 1) = 1$

(e) $6(x-2)=5x$      (f) $2(5x-3)=6x$
(g) $4(x+3)=x+16$      (h) $5(3x+2)=7x+4$
(i) $2(x+5)=4-x$      (j) $5(2x-3)=9-2x$
(k) $3(2-x)=14$      (l) $4(3-2x)=5$
(m) $6(2x-3)=4(x-2)$      (n) $4(2x-7)=3(4x-9)$
(o) $3(2x+3)=4(6-x)$      (p) $3(2x+5)=2(4x-5)+3x$
(q) $2(3x+5)-3(x-4)=1$      (r) $9x-4(2x-3)=5(x+2)$
(s) $8x-(3x-7)=2(x+3)+1$ (t) $5(3x+4)=9+(6x-7)$

## 13.5 Solving problems using equations

You can sometimes solve problems by using a letter to represent the unknown number, expressing the given information as an equation and then solving it.

### Example 13.5.1

I think of a number, multiply it by 5 and subtract 4. I get the same answer if I add 2 to the number and multiply the result by 3. Find the number.
Let $x$ be the number.

| | |
|---|---|
| The equation is | $5x-4=3(x+2)$ |
| Expand the brackets | $5x-4=3x+6$ |
| Subtract $3x$ from both sides | $2x-4=6$ |
| Add 4 to both sides | $2x=10$ |
| Divide both sides by 2 | $x=5.$ |

The number is 5.
Check in the original problem.
$5\times5=25$ and $25-4=21$; $5+2=7$ and $3\times7=21$.

### Example 13.5.2

Trevor is 8 years younger than his sister, Gwen. The sum of their ages is 110 years. Find Trevor's age.
Let Trevor's age be $x$ years. Then Gwen's age is $x+8$ years.

| | |
|---|---|
| The equation is | $x+(x+8)=110$ |
| Remove brackets | $x+x+8=110$ |
| Collect like terms | $2x+8=110$ |
| Subtract 8 from both sides | $2x=102$ |
| Divide both sides by 2 | $x=51.$ |

Trevor's age is 51 years.
Check If Trevor is 51, Gwen is $51 + 8 = 59$. The sum of their ages is $51 + 59 = 110$.

## EXERCISE 13.5

1 I think of a number, multiply it by 4 and add 7. The answer is 55. Find the number.

2 I think of a number, add 7 to it and multiply the result by 6. The answer is 90. Find the number.

3 I think of a number, multiply it by 3 and add 7. I get the same answer if I multiply the number by 4 and subtract 12. Find it.

4 I think of a number, multiply it by 7 and subtract 12. I get the same answer if I add 2 to the number and multiply the result by 5. Find the number.

5 John is 7 years older than his wife, Anne. The sum of their ages is 63 years. Find John's age.

6 Rachel is four times as old as her daughter. The sum of their ages is 40 years. Find their ages.

7 The length of a rectangle is 3 cm greater than its width. The perimeter of the rectangle is 34 cm. Find its length.

8 One of the angles of an isosceles triangle is 114°. Find the size of the other two angles.

9 The sum of three consecutive numbers is 27. Find the numbers.

10 In a soccer league, teams get 3 points for a win, 1 point for a draw and 0 points for a defeat. A team with 18 points has won 2 more games than it has drawn. How many games has it won?

# 13.6 Solving inequalities

To solve inequalities, you may use almost all the techniques you have used to solve equations. You may add the same amount to both sides, subtract the same amount from both sides and multiply or divide both sides by the same *positive* number. You can verify all of these operations by thinking of a balance where the scales do not balance.

You may *not*, however, multiply or divide both sides by the same *negative* number.

To show this, start with a true inequality such as $3 > -6$. If you multiply both sides by a negative number, say $-2$, you get $-6 > 12$, which is *false*. Similarly, if you divide both sides of $3 > -6$ by the same negative number, say $-3$, you get $-1 > 2$, which is also false.

To solve the inequality $-5x < -10$, first rewrite it as $10 < 5x$ by adding $10 + 5x$ to both sides. Now divide by 5, getting $x > 2$. You can see that the left and right sides of the original inequality $-5x < -10$ have been divided by $-5$, but the sign of the inequality has been reversed.

---

**Example 13.6.1**

Solve $3x - 4 > 14$.
Add 4 to both sides      $3x > 18$
Divide both sides by 3     $x > 6$.

---

**Example 13.6.2**

Solve $3(2x + 7) \leq 9 - 2x$.
Expand the brackets          $6x + 21 \leq 9 - 2x$
Add $2x$ to both sides         $8x + 21 \leq 9$
Subtract 21 from both sides      $8x \leq -12$
Divide both sides by 8          $x \leq -1\frac{1}{2}$.

---

**Example 13.6.3**

Solve $8 - 3x \leq 7 - x$.
Add $3x$ to both sides         $8 \leq 7 + 2x$
Subtract 7 from both sides     $1 \leq 2x$
Divide both sides by 2          $\frac{1}{2} \leq x$
Rewrite with $x$ on the left     $x \geq \frac{1}{2}$.

---

Notice how, when the inequality is rewritten to have $x$ on the left, the sign of the inequality needs to be reversed.

## EXERCISE 13.6

**1** Solve these inequalities.

(a) $4x < 12$

(b) $x + 7 \geq 2$

(c) $5x - 3 \leq 12$

(d) $7 - 2x > 1$

(e) $4x + 5 \geq 1$

(f) $3(x - 2) > 15$

(g) $2(x + 4) < 6$

(h) $x + 1 > 3x$

(i) $8x + 1 \leq 4x - 3$

(j) $4(x + 2) < 5x - 1$

(k) $2(x - 2) \geq 5(x + 1)$

(l) $5x + 1 > 10 - x$

(m) $6x - 5 \leq 15 - 2x$

(n) $4(2x + 3) < 7 - 2x$

(o) $1 - 8x \geq 11 - 4x$

# percentages

**In this chapter you will learn:**

- how to convert between percentages and fractions and decimals
- how to find percentages of quantities
- how to express one quantity as a percentage of another
- how to find percentage changes.

# 14.1 Introduction

A percentage is a fraction expressed in hundredths. The name is derived from the Latin *per centum*, which means 'per hundred'. The symbol % is often used for 'per cent'; so, for example, 19% means $\frac{19}{100}$.

Although the Romans did not use percentages, they used fractions which converted easily to hundredths to calculate taxes. From the Middle Ages, as commerce grew, so did the use of percentages. By the beginning of the seventeenth century their use was widespread, especially for the computation of interest and profit and loss.

# 14.2 Percentages, decimals, fractions

You can easily write any percentage as a fraction with 100 as its denominator, for example $37\% = \frac{37}{100}$. This fraction is already in its simplest form but, in other cases, you can simplify the fraction. For example, $48\% = \frac{48}{100}$, which, in its simplest form, is $\frac{12}{25}$.

---

**Example 14.2.1**

Write each percentage as a fraction in its simplest form.
(a) 7%　(b) 60%　(c) 200%　(d) 350%

(a) $7\% = \frac{7}{100}$.　　　　(b) $60\% = \frac{60}{100} = \frac{3}{5}$.

(c) $200\% = \frac{200}{100} = 2$.　(d) $350\% = \frac{350}{100} = 3\frac{1}{2}$.

---

When the percentage includes a fraction, for example, $17\frac{1}{2}\%$, first write the percentage as a fraction with a numerator of 100. Then multiply numerator and denominator by a number which will give you a numerator which is a whole number. Finally, if necessary, give the fraction in its simplest form.

---

**Example 14.2.2**

Write $17\frac{1}{2}\%$ as a fraction in its simplest form.

As a fraction with a numerator of 100　　　$17\frac{1}{2}\% = \dfrac{17\frac{1}{2}}{100}$

Multiply numerator and denominator by 2　$\dfrac{17\frac{1}{2} \times 2}{100 \times 2} = \dfrac{35}{200}$

In its simplest form　$\frac{35}{200} = \frac{7}{40}$.

---

You can write a percentage straight down as a decimal. For example, $23\% = \frac{23}{100}$ which, as a decimal, is 0.23. (You might find it helpful to think of $\frac{23}{100}$ as $23 \div 100$).

---

**Example 14.2.3**

Write each percentage as a decimal.

(a) 47%  (b) 3%  (c) 60%  (d) 213%  (e) 120%
(f) 7.8%  (g) $23\frac{1}{2}\%$

(a) $47\% = 0.47$          (b) $3\% = 0.03$
(c) $60\% = 0.60 = 0.6$    (d) $213\% = 2.13$
(e) $120\% = 1.20 = 1.2$   (f) $7.8\% = \frac{7.8}{100} = 7.8 \div 100 = 0.078$
(g) $23\frac{1}{2}\% = 23.5\% = 0.235$

---

Reversing this process, you can write a decimal as a percentage.

---

**Example 14.2.4**

Write each decimal as a percentage.

(a) 0.39  (b) 0.07  (c) 0.9  (d) 1.08  (e) 0.475
(f) 0.043

(a) $0.39 = 39\%$          (b) $0.07 = 7\%$
(c) $0.9 = 90\%$           (d) $1.08 = 108\%$
(e) $0.475 = 47.5\%$       (f) $0.043 = 4.3\%$

---

You can write a fraction with a denominator of 100 straight down as a percentage. For example, $\frac{29}{100} = 29\%$ . For a fraction with a denominator which is a factor or a multiple of 100, convert it to an equivalent fraction with a denominator of 100 and then express it as a percentage.

---

**Example 14.2.5**

Express each fraction as a percentage.

(a) $\frac{7}{25}$  (b) $\frac{135}{500}$

(a) Multiply numerator and denominator by 4:
   $\frac{7}{25} = \frac{28}{100} = 28\%$.

(b) Divide numerator and denominator by 5:
   $\frac{135}{500} = \frac{27}{100} = 27\%$.

---

Alternatively, to convert any fraction to a percentage, think of the line in the fraction as a division sign and then, by hand or with a calculator, work out the result as a decimal, and then convert it to a percentage.

---

**Example 14.2.6**

Express each fraction as a percentage.
(a) $\frac{3}{4}$  (b) $\frac{7}{8}$

(a) $\frac{3}{4} = 3 \div 4 = 0.75 = 75\%$.  (b) $\frac{7}{8} = 7 \div 8 = 0.875 = 87.5\%$.

---

### EXERCISE 14.1

1 Write each percentage as a fraction in its simplest form.
   (a) 49%    (b) 50%    (c) 25%    (d) 10%
   (e) 65%    (f) 1%     (g) 3%     (h) 100%
   (i) 250%   (j) $12\frac{1}{2}\%$   (k) $33\frac{1}{3}\%$   (l) $13\frac{3}{4}\%$

2 Write each percentage as a decimal.
   (a) 39%    (b) 7%     (c) 30%    (d) 112%
   (e) 3.8%   (f) $62\frac{1}{2}\%$   (g) $3\frac{1}{2}\%$   (h) $12\frac{1}{4}\%$

3 Write each decimal as a percentage.
   (a) 0.71   (b) 0.04   (c) 0.8    (d) 1.75
   (e) 1.9    (f) 0.027  (g) 0.0325 (h) 0.009

4 Write each fraction as a percentage.
   (a) $\frac{43}{100}$   (b) $\frac{9}{100}$   (c) $\frac{7}{10}$   (d) $\frac{31}{50}$
   (e) $\frac{17}{25}$   (f) $\frac{4}{5}$   (g) $\frac{165}{300}$   (h) $\frac{3}{8}$

5 Approximately 52% of babies born are boys.
   (a) Write 52%
       (i) as a decimal    (ii) as a fraction.
   (b) What percentage of babies born are girls?

6 Approximately 45% of people who take their car driving test pass.
   (a) What percentage fail?    (b) What fraction fail?

## 14.3 Percentages of quantities

You can find a percentage of a quantity mentally when the percentage has a well-known fractional equivalent and the arithmetic is simple. For example, to find 50% of 18, use the fact that $50\% = \frac{1}{2}$ and then find $\frac{1}{2}$ of 18; in other words $\frac{1}{2} \times 18 = 9$.

---

**Example 14.3.1**

Find 75% of 24 km.

$75\% = \frac{3}{4}$. So $\frac{3}{4}$ of $24 = \frac{3}{4} \times 24 = 18$.

---

In general, to find a percentage of a quantity, write the percentage as a fraction or as a decimal; replace 'of' by '×' and carry out the multiplication, either by hand or using a calculator.

---

**Example 14.3.2**

Find 36% of 175.

As a fraction $\qquad 36\% = \frac{36}{100}$

Replace 'of' by '×' $\quad 36\%$ of $175 = \frac{36}{100} \times 175 = 63$

---

**Example 14.3.3**

Find 47% of 260 litres.

As a decimal $\qquad 47\% = 0.47$

Replace 'of' by '×' $\qquad 47\%$ of $260 = 0.47 \times 260 = 122.2$

Therefore 47% of 260 litres is 122.2 litres.

---

When you find a percentage of an amount of money, you sometimes have to round your answer to the nearest penny.

---

**Example 14.3.4**

Find 37% of £6.53 and give your answer to the nearest penny.

$0.37 \times 6.53 = 2.4161$, so 37% of £6.53 = £2.4161.

To the nearest penny, 37% of £6.53 = £2.42.

---

1  Without using a calculator, find
   (a)  50% of 36      (b)  25% of 48
   (c)  75% of 36 kg   (d)  10% of 250 m
   (e)  20% of 45 cm   (f)  $33\frac{1}{3}$% of 27 litres

2  Using a calculator if necessary, find
   (a)  30% of 80              (b)  46% of 250
   (c)  76% of 32              (d)  8% of 475 m
   (e)  12% of 540             (f)  120% of 85
   (g)  250% of 386            (h)  115% of 680 kg
   (i)  15.6% of 800 litres    (j)  9.3% of 30 km
   (k)  $27\frac{1}{2}$% of 72 cm   (l)  $7\frac{1}{2}$% of 620 m

3  There are 1200 students in a school. 57% of them are girls.
   (a)  Work out the number of girls.
   (b)  What percentage of the students are boys?
   (c)  Work out the number of boys.

4  A test was marked out of 80. Mary scored 85% of the marks.
   How many marks did she score?

5  Find 55% of 1 day and give your answer in minutes.

6  Find
   (a)  92% of £865     (b)  78% of £426     (c)  45% of £364

7  Find 19% of £9.36 and give your answer to the nearest penny.

8  David pays a deposit of 15% on a car which costs £8200.
   (a)  How much deposit does he pay?
   (b)  What percentage of the cost has he still to pay?

# 14.4  Increasing and decreasing quantities

Sometimes you have to find a percentage of a quantity and then add it to or subtract it from the original amount. For example, if you put £600 in a bank account and receive 5% interest after one year, the amount in your account after one year is not 5% of £600 (£30). You must add the £30 to the original £600. There is £630 in your account.

### Example 14.4.1

(a) Increase 350 by 6%.   (b) Decrease 480 by 5%.

(a) 6% of $350 = 0.06 \times 350 = 21$. The new amount is $350 + 21 = 371$.

(b) 5% of $480 = 0.05 \times 480 = 24$. The new amount is $480 - 24 = 456$.

### EXERCISE 14.3

1  Increase
   (a) 65 by 20%,   (b) 400 by 9%,   (c) 840 by 4%,
   (d) 5600 by 14%,   (e) 70 by 12%,   (f) 750 by 3%.

2  Decrease
   (a) 98 by 50%,   (b) 430 by 10%,   (c) 240 by 5%,
   (d) 3200 by 15%,   (e) 820 by 24%,   (f) 45 by 18%.

3  The number of people employed by a company increased by 16% from 650. How many people are employed by the company now?

4  The normal price of a computer is £550. In a sale, normal prices are reduced by 15%. Find the sale price of the computer.

5  Julie's salary was £2350 per month before she received a pay rise of 4%. Find her new monthly salary.

6  A year ago, Richard bought a new car for £12 680. Since then, it has lost 12% of its value. Find his car's current value.

7  Chris invests £630 in a bank account at an interest rate of 4.5% per annum. (Per annum means per year.) After one year, the interest is added to her account. How much is in her account after one year?

8  Tony buys a camera for £130 and later sells it, making a loss of 12.5%. Find the selling price.

## 14.5  One quantity as a percentage of another quantity

To express one quantity as a percentage of another quantity, first express it as a fraction of the other quantity. Then convert the fraction to a decimal and finally to a percentage.

**Example 14.5.1**

Elaine scores 52 out of 80 in a test. Express her mark as a percentage.

As a fraction, 52 out of 80 is $\frac{52}{80}$.

Find $52 \div 80$ to convert $\frac{52}{80}$ to a decimal, $\frac{52}{80} = 0.65$.

As a percentage $0.65 = 65\%$.

If both quantities are expressed in the same units, the method is the same as if both quantities were just numbers.

If the quantities are expressed in different units, however, you must express them in the same units, usually the smaller of the two.

**Example 14.5.2**

(a) Express $84\,m$ as a percentage of $230\,m$. Give your answer correct to one decimal place.

(b) Express $39\,mm$ as a percentage of $15\,cm$.

(a) As a fraction, you require $\frac{84}{230}$.
  As a decimal, this is $84 \div 230 = 0.3652\ldots$.
  As a percentage, this is $36.5\%$ (correct to 1 d.p.).

(b) In the same units, the quantities are $39\,mm$ and $150\,mm$.
  As a fraction, you require $\frac{39}{150}$.
  As a decimal, this is $39 \div 150 = 0.26$.
  As a percentage, this is $26\%$.

**EXERCISE 14.4**

1 Tony scores 63 out of 75 in a test. Express this as a percentage.

2 In a class of 35 students, 21 are girls. What percentage of the students are girls and what percentage are boys?

3 A team played 38 matches and won 21. What percentage of its matches did it win? Give your answer correct to one decimal place.

4 The area of the surface of the Earth is $510\,066\,000\,km^2$, of which $361\,743\,000\,km^2$ is water. What percentage of the

Earth's surface is water? Give your answer correct to the nearest whole number.

5 Express the first quantity as a percentage of the second. Where necessary, give your answer correct to one decimal place.
(a) 9 cm, 25 cm     (b) 24 m, 60 m     (c) 18 hrs, 24 hrs
(d) 29 kg, 82 kg     (e) 18 mm, 4 cm     (f) 85 cm, 5 m
(g) 14 hrs, 3 days     (h) 350 m, 3 km     (i) 83 p, £7

# 14.6 Percentage increases

When a quantity changes, you can find the percentage increase in a quantity by expressing the change as a percentage of the *original* quantity. The same applies to decreases.

---

**Example 14.6.1**

The population of a village increases from 675 to 783. Find the percentage increase.

The increase is $783 - 675 = 108$, on an original total of 675.

As a percentage, this is $\dfrac{\text{increase}}{\text{original}} = \dfrac{108}{675} = 0.16 = 16\%$.

The percentage increase is 16%.

---

**Percentage profit** and **percentage loss** are important types of percentage change. For example, if you buy a car for £8000 (the cost price) and sell it for £6800 (the selling price), you make a loss of $£(8000 - 6800) = £1200$. To find the percentage loss, express the loss as a percentage of the cost price, $\frac{1200}{8000} = 0.15 = 15\%$.

---

**Example 14.6.2**

Joan bought a house for £62 500 and later sold it for £75 000. Find her percentage profit.

$\text{Profit} = £(75\,000 - 62\,500) = £12\,500$

As a percentage, this is $\dfrac{\text{profit}}{\text{cost price}} = \dfrac{12\,500}{62\,500} = 0.2 = 20\%$.

The percentage profit is 20%.

---

## EXERCISE 14.5

1 The number of workers employed by a company increased from 225 to 243. Find the percentage increase.

2 The population of a town fell from 23 650 to 20 812. Find the percentage decrease.

3 Helen's annual salary was increased from £22 500 to £23 400. Find her percentage increase.

4 The original price of a bicycle was £184. In a sale, its price was £161. Find the percentage reduction.

5 Bill put £5500 in a bank account. A year later, interest was added and the amount in his account increased to £5885. Find the rate of interest.

6 The cost price of a pair of shoes was £24 and the selling price was £42. Find the percentage profit.

7 Kim bought a computer for £750 and later sold it for £495. Find her percentage loss.

8 An art dealer buys a painting for £3500 and sells it for £4480. Find his percentage profit.

# 14.7  Using multipliers

If normal prices are reduced by 15% in a sale, the sale price of an item is 85%, that is, (100% − 15%) of its normal price. Thus, to find the sale price of a television, the normal price of which is £360, you find 85% of £360. As 0.85 × 360 = 306, the sale price is £306.

Similarly, if the number of employees in a company increases by 12%, the new number is 112%, that is, (100% + 12%) of the original number. So, if the number of employees was originally 350, the new number is 112% of 350. As 112% of 350 = 1.12 × 350 = 392, the new number of employees is 392.

The numbers 0.85 and 1.12 are called **multipliers**.

---

**Example 14.7.1**

Write down the multipliers needed for these changes.
(a) a decrease of 23%,   (b) an increase of 20%,
(c) a decrease of 9%.

(a) $100\% - 23\% = 77\%$, so the multiplier is 0.77.
(b) $100\% + 20\% = 120\%$, so the multiplier is 1.20 or 1.2.
(c) $100\% - 9\% = 91\%$, so the multiplier is 0.91.

Multipliers are especially useful for calculating **compound interest,** the type of interest paid by banks and building societies. Compound interest is paid on the *total* of the amount invested (called the **principal**) and the interest earned in previous years.

For example, if you invest £5000 at 6% p.a. compound interest, the amount in your account after one year will be $5000 \times 1.06$, that is, £5300. After two years, the new amount increases to $5000 \times 1.06 \times 1.06 = 5000 \times 1.06^2 = 5618$, that is, £5618. After three years, the amount will be $£5000 \times 1.06^3 = £5955.08$, and so on.

---

**Example 14.7.2**

Jan invests £10,000 at 4.5% p.a. compound interest. Work out
(a) the amount in her account after 3 years,
(b) the amount of interest earned.
(c) $10\,000 \times 1.045^3 = 11\,411.661$, so the amount in her account is £11411.66, to the nearest penny.
(d) Amount of interest earned $= £11411.66 - £10000$
$= £1411.66$

---

You can also use multipliers to find the original quantity *before* a percentage change, if you know the percentage change and the new quantity *after* the change.

For example, if normal prices are reduced by 12% in a sale, you multiply the normal price of an item by 0.88 to find its sale price.

Thus, normal price $\times 0.88 =$ sale price.

Working back, if you know the sale price, divide it by 0.88 (the inverse operation to multiplying by 0.88) to find the normal price. If the sale price is £396, find $396 \div 0.88 = 450$, giving a normal price of £450.

**Example 14.7.3**

The cost of a airline ticket is increased by 14%. After the increase, the cost of the ticket is £399. Find its cost before the increase.

As

cost before increase × 1.14 = cost after increase

so

cost before increase = cost after increase ÷ 1.14.

As  399 ÷ 1.14 = 350,  the ticket cost £350 before the increase.

## EXERCISE 14.6

1 Write down the multiplier needed for each of these changes.
   (a) a decrease of 31%        (b) an increase of 43%
   (c) a decrease of 30%        (d) an increase of 7%
   (e) an increase of 17.5%     (f) a decrease of 3.4%

2 Normal prices are reduced by 35% in a sale. The normal price of a video recorder is £260. Find its sale price.

3 The population of the United Kingdom was 50.3 million in 1951. By 1991, it had increased by 15%. Find the population in 1991. Give your answer correct to the nearest 0.1 million.

4 Joe invests £8000 at 7% p.a. compound interest. Work out the amount in his account after 2 years.

5 Jean invests £500 at 3.5% p.a. compound interest. Work out
   (a) the amount in her account after 3 years,
   (b) the amount of interest earned.

6 Janet invests a sum of money at an interest rate of 8% p.a. After one year, she has £810 in her account. How much did she invest?

7 In a sale, normal prices are reduced by 20%. The sale price of a microwave oven is £140. Find its normal price.

8 The price of a new car is increased by 18% to £14 750. Find its cost before the increase.

# 15

## formulae

In this chapter you will learn:

- how to write formulae
- how to evaluate and use formulae
- how to change the subject of a formula.

# 15.1 What is a formula?

In mathematics and science, a **formula** is a quick way of expressing a rule or a fact. You have already met some formulae (the plural of formula) expressed in words, area of parallelogram = base × height, for example, in Chapter 10. If you use $A$ to stand for the area of a parallelogram, $b$ to stand for its base and $h$ to stand for its height, you can write this result more concisely as the formula $A = bh$.

$A$ is called the **subject** of this formula, because it appears on its own on the left-hand side and does not appear on the right-hand side.

A multiplication sign is not necessary in the formula $A = bh$. This is one of the many ideas from Chapter 11 you will need in this chapter. You will, for example, evaluate formulae in exactly the same ways as you evaluated algebraic expressions. A formula is not, however, the same as an expression; for example, a formula has an equals sign in it, whereas an expression does not.

---

**Example 15.1.1**

Each side of a square is $l$ centimetres long. The perimeter of the square is $P$ centimetres.

(a) Write down a formula for $P$ in terms of $l$. Give your answer as simply as possible.
(b) Evaluate $P$ when $l = 3.7$.
(a) Perimeter of a square = 4 × length of each side, so $P = 4l$.
(b) When $l = 3.7$, $P = 4 \times 3.7 = 14.8$

---

### EXERCISE 15.1

1 Area of a triangle $= \frac{1}{2} \times$ base × height.
 (a) Write down a formula for the area, $A$, of a triangle in terms of its base, $b$, and its height $h$.
 (b) Evaluate $A$ when $b = 7$ and $h = 8$.

2 Average speed = distance travelled ÷ time taken.
 (a) Write down a formula for the average speed, $s$ miles per hour, of a motorist who travels $d$ miles in $t$ hours.
 (b) Evaluate $s$ when $d = 126$ and $t = 2.25$.

3 To find the angle sum, in degrees, of a polygon, subtract 2 from the number of sides it has and then multiply the result by 180.
  (a) Find the angle sum, $S$, of a polygon with $n$ sides.
  (b) Evaluate $S$ when $n = 9$.

4 Each side of a regular octagon is $d$ centimetres long. The perimeter of the octagon is $P$ centimetres.
  (a) Write down a formula for $P$ in terms of $d$.
  (b) Evaluate $P$ when $d = 3.7$.

5 Each side of a square is $l$ centimetres long. The area of the square is $A$ square centimetres.
  (a) Write down a formula for $A$ in terms of $l$.
  (b) Evaluate $A$ when $l = 8.6$.

6 $a$, $b$ and $c$ are three numbers. The mean of the numbers is $m$.
  (a) Write down a formula for $m$ in terms of $a$, $b$ and $c$.
  (b) Evaluate $m$ when $a = 27$, $b = 41$ and $c = 37$.

7 The lengths of the parallel sides of a trapezium are $a$ centimetres and $b$ centimetres. The parallel sides are $h$ centimetres apart. The area of the trapezium is $A$ square centimetres.
  (a) Write down a formula for $A$ in terms of $a$, $b$ and $h$.
  (b) Evaluate $A$ when $a = 12.6$, $b = 7.4$ and $h = 5.8$.

8 A regular polygon has $n$ sides. The size of each of its exterior angles is $x°$.
  (a) Write down a formula for $x$ in terms of $n$.
  (b) Evaluate $x$ when $n = 9$.

9 $T = \frac{1}{2}n(n+1)$. Evaluate $T$ when $n = 10$. (This value is the tenth triangle number.)

10 $v = u + at$. Evaluate $v$ when $u = 20$, $a = -3$ and $t = 4$.

11 $s = \frac{1}{2}gt^2$. Evaluate $s$ when $g = -10$ and $t = 6$.

12 $I = \frac{1}{100} PRT$. Evaluate $I$ when $P = 1200$, $R = 3\frac{1}{2}$ and $T = 8$.

## 15.2 Evaluating terms other than the subject

In each question of Exercise 15.1, the term you had to evaluate was the subject of a formula. Sometimes you will have to find the value of a term which is not the subject. To do this, you substitute into the formula the values you are given and then solve the resulting equation.

If, for example, in the formula $A = bh$, you are told that $A = 54$ and $b = 6$, substituting these values gives the equation $6h = 54$, which has the solution $h = 9$.

**Example 15.2.1**

$v = u + at$. Given that $u = 6$, $v = 38$ and $a = 4$, find the value of $t$.

Substituting the values of $u$, $v$ and $a$ gives $38 = 6 + 4t$.

Subtracting 6 from both sides gives $4t = 32$.

Dividing both sides by 4 gives $t = 8$.

**Example 15.2.2**

In the formula $a = kv^2$, $a = 45$ and $k = 5$. Find the value of $v$.

Substituting the values of $a$ and $k$ gives $45 = 5v^2$.

Dividing both sides by 5 gives $v^2 = 9$.

Therefore $v = 3$ or $v = -3$.

The number 3 is the **square root** of 9, written $\sqrt{9}$, because $3^2 = 9$, and is positive.

In general, $x$ is the square root of $a$, written $\sqrt{a}$, if $x^2 = a$ and $x \geq 0$.

### EXERCISE 15.2

1   $P = 3d$. Find the value of $d$ when $P = 42$.

2   $A = lb$. Find the value of $b$ when $A = 108$ and $l = 9$.

3   $y = 4x - 3$. Find the value of $x$ when $y = 7$.

4   $P = 2x + y$. Find the value of $y$ when $P = 23$ and $x = 8$.

5   $P = 2l + 2b$. Find the value of $b$ when $P = 28$ and $l = 8$.

6   $I = \frac{1}{100}PRT$. Find the value of $R$ when $I = 360$, $P = 1500$ and $T = 6$.

7   $s = \dfrac{d}{t}$. Find the value of $d$ when $s = 84$ and $t = 1.5$.

8  $v = u + at$. Find the value of $a$ when $v = 13$, $u = 25$ and $t = 4$.

9  $P = 2(l + b)$. Find the value of $l$ when $P = 16$ and $b = 3$.

10  $s = ut + \frac{1}{2}at^2$. Find the value of $u$ when $s = 52$, $t = 4$ and $a = 5$.

11  $a = rw^2$. Find the value of $w$ when $a = 75$ and $r = 3$.

12  $s = \frac{1}{2}gt^2$. Find the value of $t$ when $s = 80$ and $g = 10$.

13  $P = \dfrac{V^2}{R}$. Find the value of $V$ when $P = 20$ and $R = 5$.

14  $F = \dfrac{mv^2}{r}$. Find the value of $v$ when $F = 24$, $m = 8$ and $r = 12$.

## 15.3  Changing the subject of a formula

The techniques needed to change the subject of a formula are the same as those you used to solve equations in Chapter 13. Formulae are really equations. You must do the same to both sides of a formula, just as you did the same to both sides of an equation.

For example, to make $l$ the subject of the formula $A = lb$, divide both sides by $b$, obtaining $\dfrac{A}{b} = l$, which is $l = \dfrac{A}{b}$. Similarly, to make $b$ the subject, divide both sides by $l$, obtaining $b = \dfrac{A}{l}$.

---

**Example 15.3.1**

Make $P$ the subject of the formula $\dfrac{PV}{T} = k$.

Multiply both sides by $T$     $PV = kT$.

Divide both sides by $V$     $P = \dfrac{kT}{V}$.

---

**Example 15.3.2**

Make $h$ the subject of the formula $A = \frac{1}{2}bh$.

Multiply both sides by 2     $2A = bh$

Divide both sides by $b$     $h = \dfrac{2A}{b}$.

**Example 15.3.3**

Make $t$ the subject of the formula $v = u + at$.

Subtract $u$ from both sides $\quad v - u = at$

Divide both sides by $a$ $\quad t = \dfrac{v - u}{a}$.

**Example 15.3.4**

Make $u$ the subject of the formula $I = m(v - u)$.

Expand the brackets $\quad I = mv - mu$

Add $mu$ to both sides $\quad mu + I = mv$

Subtract $I$ from both sides $\quad mu = mv - I$

Divide both sides by $m$ $\quad u = \dfrac{mv - I}{m}$.

## EXERCISE 15.3

1 In each part, make the letter in brackets the subject of the formula.

(a) $V = IR$ $\quad$ (R) $\qquad$ (b) $V = bh$ $\quad$ (b)

(c) $E = mgh$ $\quad$ (h) $\qquad$ (d) $P = 2x + y$ $\quad$ (y)

(e) $S = \dfrac{D}{T}$ $\quad$ (T) $\qquad$ (f) $V = \frac{1}{3}Ah$ $\quad$ (h)

(g) $I = \dfrac{PRT}{100}$ $\quad$ (R) $\qquad$ (h) $\dfrac{PV}{T} = k$ $\quad$ (V)

(i) $p = a + b + c$ $\quad$ (a) $\qquad$ (j) $P = 2x + y$ $\quad$ (x)

(k) $v = u - gt$ $\quad$ (u) $\qquad$ (l) $v = u + at$ $\quad$ (a)

(m) $v = u - gt$ $\quad$ (t) $\qquad$ (n) $P = 2(l + b)$ $\quad$ (l)

(o) $I = m(v - u)$ $\quad$ (v) $\qquad$ (p) $S = 180n - 360$ $\quad$ (n)

(q) $2S = a + b + c$ $\quad$ (a) $\qquad$ (r) $A = \dfrac{h(a + b)}{2}$ $\quad$ (a)

# 16

## circles

**In this chapter you will learn:**

- the names of parts of a circle
- how to find the circumference of a circle
- how to find the area of a circle
- how to use two circle properties.

# 16.1 Introduction

A **circle** is a plane curve, every point of which is the same distance from a fixed point, called the **centre** of the circle. This distance from the centre to a point on the circle is the **radius**. (See Figure 16.1.)

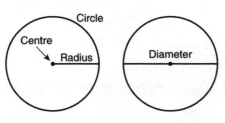

**Figure 16.1**        **Figure 16.2**

A straight line which joins two points on the circle and passes through the centre is a **diameter.** (See Figure 16.2.) So a diameter divides a circle exactly in half, or, the diameter **bisects** the circle. This was one of the fundamental propositions of Thales (c 600 BC), on which classical geometry was based.

A diameter divides a circle into two **semicircles** and is a **line of symmetry.** As you can draw an infinite number of diameters on a circle, you can also draw an infinite number of lines of symmetry. Similarly, the **order of rotational** symmetry of a circle is also infinite, as a tracing of a circle can be rotated to fit on top of itself in an infinite number of different positions.

A diameter is twice as long as a radius. If you use $d$ to stand for the diameter and $r$ to stand for the radius, then $d = 2r$.

An **arc** is a part of a circle. The word 'arc' comes from the Latin for 'bow'. The shaded area in Figure 16.3 (overleaf) formed by an arc and two radii is a **sector.** You met this term in Chapter 06 as the mathematical name for the 'slice' of a pie chart.

A straight line joining two points on a circle is a **chord.** A diameter is, therefore, a special type of chord.

The shaded region in Figure 16.4 (overleaf) formed by an arc and a chord is called a **segment.** The straight line in Figure 16.5 which just touches a circle is a **tangent,** from the Latin for 'touch'. A line is a tangent to a circle if, when extended in either direction, it meets the circle only once.

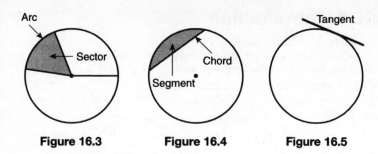

**Figure 16.3**      **Figure 16.4**      **Figure 16.5**

## 16.2 Circumference of a circle

**Circumference** is the special name of the perimeter of a circle, that is, the distance all around it. The word is derived from the Latin *circum*, meaning 'round' and *fero*, meaning 'carry'.

If you measure the circumferences and diameters of some circular objects and then, for each one, work out the value of $\frac{\text{circumference}}{\text{diameter}}$, you should find that your answer is always just over 3. In fact, the value of $\frac{\text{circumference}}{\text{diameter}}$ is the same for every circle and is denoted by the Greek letter $\pi$, pronounced pi. So, for every circle $\frac{\text{circumference}}{\text{diameter}} = \pi$.

William Jones, an Englishman, was, in 1706, the first mathematician to use $\pi$ in this way, probably because $\pi$ is the first letter of the Greek word *peripheria*, meaning 'periphery' or 'boundary'.

Archimedes (c 225 BC) showed that the value of $\pi$ is between $3\frac{1}{7}$ and $3\frac{10}{71}$. Using modern computing techniques, its value has been calculated to over six billion decimal places but it cannot be found exactly, as the decimal never recurs or terminates.

To perform calculations involving $\pi$, use the $\pi$ key on your calculator. If you do not have a $\pi$ key, 3.14 or 3.142 is sufficiently accurate for all practical purposes.

Using $C$ to stand for the circumference of a circle and $d$ to stand for its diameter, you can write the result $\frac{\text{circumference}}{\text{diameter}} = \pi$ as $\frac{C}{d} = \pi$. Multiplying both sides by $d$, you obtain the formula $C = \pi d$. So, to find the circumference of a circle, you multiply its diameter by $\pi$. The same units must be used for the circumference and the diameter.

**Example 16.2.1**

The diameter of a circle is 6.3 cm. Calculate its circumference. Give your answer correct to three significant figures.

If $C$ cm is the circumference, then $C = \pi \times 6.3 = 19.8$, so the circumference is 19.8 cm, correct to three significant figures.

---

If you are told the radius of a circle and want to find its circumference, you could first double the radius to obtain the diameter and then multiply the result by $\pi$.

Alternatively, using $r$ to stand for the radius of the circle, you can replace $d$ by $2r$ in the formula $C = \pi d$. This gives $C = \pi \times 2r$, usually written as $C = 2\pi r$. In this formula, the same units must be used for the circumference and the radius.

**Example 16.2.2**

The radius of a circle is 1.96 m. Calculate its circumference. Give your answer correct to three significant figures.

If $C$ m is the circumference, then $C = 2 \times \pi \times 1.96 = 12.3$, so the circumference is 12.3 m, correct to three significant figures.

---

If you know the circumference of a circle and want to find its diameter, you could substitute the value of $C$ in the formula $C = \pi d$ and then solve the resulting equation.

Alternatively, you can make $d$ the subject of the formula $C = \pi d$.

Dividing both sides by $\pi$ gives $d = \dfrac{C}{\pi}$. So, to find the diameter of a circle, you divide its circumference by $\pi$.

**Example 16.2.3**

The circumference of a circle is 57.8 cm. Calculate its diameter. Give your answer correct to three significant figures.

As $d = \dfrac{C}{\pi}$, $d = \dfrac{57.8}{\pi} = 18.4$. The diameter is 18.4 cm, correct to three significant figures.

If you know the circumference of a circle and want to find its radius, you could find the diameter of the circle and then halve your answer.

Alternatively, you could make $r$ the subject of the formula $C = 2\pi r$.

Dividing both sides of the formula by $2\pi$ gives $r = \dfrac{C}{2\pi}$. So, to find the radius of a circle, you divide its circumference by $2\pi$.

### EXERCISE 16.1

Give all answers correct to three significant figures.

1  The diameter of a circle is 3.7 cm. Calculate its circumference.

2  The radius of a circle is 6.34 m. Calculate its circumference.

3  The circumference of a circle is 76 cm. Calculate its diameter.

4  The circumference of a circle is 97 m. Calculate its radius.

5  The diameter of the Earth at the equator is 12 756 km. Calculate the circumference of the Earth to the nearest hundred km.

6  The minute hand of Big Ben is 3.35 m long. How far does the tip of this hand move in 30 minutes?

7  The diameter of a bicycle wheel is 71 cm.
  (a)  Calculate the circumference of the wheel.
  (b)  Calculate the number of revolutions the wheel makes when the bicycle travels 1 kilometre.

8  The largest Big Wheel in the world has a circumference of 200 m. Calculate its radius.

## 16.3  Area of a circle

One of the most famous problems of antiquity was to construct, using only compasses and a straight edge, a square whose area is equal to that of a given circle. In fact, the task is impossible, precisely what is meant today when someone is said to be trying to 'square the circle'.

In attempting this problem, however, the Greeks developed methods of finding the area of a circle. One of these was the 'method of exhaustion', in which a regular polygon was drawn inside the circle and its area calculated. The more sides the regular polygon has, the nearer its area is to the circle; the area between the polygon and the circle is 'exhausted'. (See Figure 16.6 overleaf.)

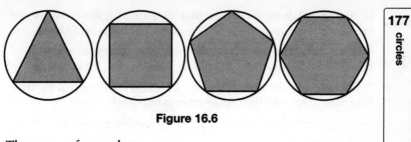

**Figure 16.6**

The area of a polygon was found by splitting it up into triangles by drawing lines from the centre of the circle to each vertex of the polygon. The area of each triangle, and thus the area of the polygon, were calculated. Figure 16.7 shows a regular octagon with perimeter $P$. It has been split into eight triangles with vertical height $h$. The length of the base of each triangle is $\frac{1}{8}P$. The area of each triangle is $\frac{1}{2} \times \frac{1}{8}P \times h$.

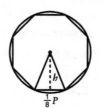

**Figure 16.7**

So the area of the octagon is $8 \times \frac{1}{2} \times \frac{1}{8}P \times h = \frac{1}{2}Ph$ or, in words, area of polygon $= \frac{1}{2} \times$ its perimeter $\times$ vertical height of triangle.

The same result is true for any regular polygon. As the number of sides of the polygon increases, the polygon becomes more and more like a circle. Its perimeter is almost equal to the circumference of the circle and the vertical height of each triangle is almost equal to the radius. Thus,

**area of a circle $= \frac{1}{2} \times$ circumference $\times$ radius.**

Using $A$ for the area of a circle, $r$ for its radius and $2\pi r$ for the circumference, you get $A = \frac{1}{2} \times 2\pi r \times r$. This simplifies to $A = \pi r^2$.

---

**Example 16.3.1**

The radius of a circle is 6.7 cm. Calculate its area. Give your answer correct to three significant figures.

If $A$ cm$^2$ is the area, $A = \pi \times 6.7^2 = 141$, so the area of the circle is 141 cm$^2$, correct to three significant figures.

It is important to recognize that the area of the circle *is not* $(\pi \times 6.7)^2$; you must find $6.7^2$ first, and then multiply by $\pi$.

You can avoid major errors like this by making an estimate of the answer. In this case, $\pi \times 6.7^2 \approx 3 \times 7^2 = 3 \times 49 \approx 150$.

If you are told the diameter of a circle and want to find its area, you must first find the radius by halving the diameter.

---

**Example 16.3.2**

The diameter of a circle is $9.56$ m. Calculate its area. Give your answer correct to three significant figures.

Let $r$ m be the radius and $A$ m$^2$ be the area.

Then $r = \frac{1}{2} \times 9.56 = 4.78$, and $A = \pi \times 4.78^2 = 71.8$. So the area of the circle is $71.8$ m$^2$, correct to three significant figures.

---

If you know the area of a circle and want to find its radius, you substitute the value of $A$ in $A = \pi r^2$ and solve the resulting equation.

---

**Example 16.3.3**

The area of a circle is $70$ cm$^2$. Calculate its radius. Give your answer correct to three significant figures.

In $A = \pi r^2$, substituting $70$ for $A$ gives $70 = \pi r^2$.

Dividing both sides by $\pi$ gives $r^2 = \dfrac{70}{\pi} = 22.28\ldots$.

Then $r = \sqrt{22.28\ldots} = 4.72$, so the radius is $4.72$ cm, correct to three significant figures.

---

Note that $r = -\sqrt{22.28\ldots}$ also solves the equation $70 = \pi r^2$, but the radius of a circle must be positive.

## EXERCISE 16.2

Give all answers correct to three significant figures.

1  The radius of a circle is $26.3$ cm. Calculate its area.

2  The diameter of a circle is $28$ m. Calculate its area.

3  The area of a circle is $100$ cm$^2$. Calculate its radius.

4  The area of a circle is $58$ cm$^2$. Calculate its diameter.

5 The radius of a semicircle is 7.31 m. Calculate its area.

6 The sides of a square piece of metal are 22 mm long. A hole with a diameter of 8 mm is drilled in the piece of metal. Calculate the area of metal which remains.

7 Two concentric circles (circles having the same centre) have radii of 4.7 cm and 6.9 cm. Calculate the area of the ring between the circles. (It is called an *annulus*.)

8 The diameter of a circular pond is 8 m. The pond has a path, which is 1 m wide, around it. Calculate the area of the path.

9 The radius of a circle is 10 cm. How many times greater is its area when its radius is doubled?

10 A farmer has 200 m of fencing. He arranges it in a circle. Calculate the area he encloses.

# 16.4  Two properties of circles

In Figure 16.8, *AB* is a chord of a circle centre *O*.

The mid-point of *AB* is *M*. The line *OM* is a line of symmetry so it is perpendicular to the chord *AB*.

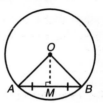

**Figure 16.8**

In general, **the line joining the centre of a circle to the mid-point of a chord is perpendicular to the chord.**

The *converse*, or opposite, of this is also true: **a line drawn from the centre of a circle perpendicular to a chord bisects the chord.**

---

**Example 16.4.1**

In the figure, *O* is the centre of the circle and *R* is the mid-point of the chord *PQ*. Angle *OPR* is 54°. Find the size of the angle marked *x*.

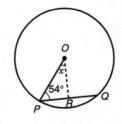

Angle $PRO = 90°$ (*OR* is perpendicular to *PQ*) $x + 90° + 54° = 180°$, (angles in a triangle), so $x + 144° = 180°$, giving $x = 36°$.

Alternatively, you could find angle *POQ* which is $180° - 2 \times 54°$ $= 72°$, and then use the fact that *OR* is a line of symmetry. *OR* therefore bisects angle *POQ*, and it follows that angle $POQ = \frac{1}{2} \times 72° = 36°$.

In Figure 16.9, *O* is the centre of a circle. A tangent touches the circle at *T*. The radius, *OT*, is a line of symmetry, so it is perpendicular to the tangent.

**In general, a radius drawn to the point where a tangent touches a circle is perpendicular to the tangent.**

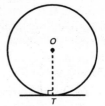

**Figure 16.9**

---

**Example 16.4.2**

In the diagram, *PQ* is the tangent at *T* to a circle centre *O*. Angle $PTR = 47°$. Calculate the size of the angle marked *y*.

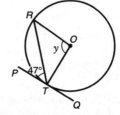

Angle $PTO = 90°$ (radius is perpendicular to tangent).

Angle $RTO = 90° - 47° = 43°$.

Angle $TRO = 43°$ (triangle *ORT* is isosceles) $43° + 43° + y = 180°$ (angle sum of triangle) giving $y + 86° = 180°$, so $y = 180° - 86° = 94°$.

---

**EXERCISE 16.3**

Find the size of each of the angles marked with letters. *O* is the centre of the circle in every question, and in Questions 3–6 a tangent is drawn to the circle.

1

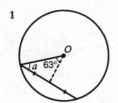

2

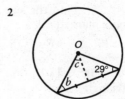

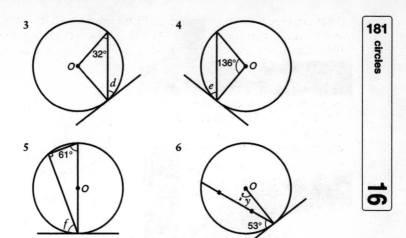

3

32°

d

4

136°

e

5

61°

f

6

y

53°

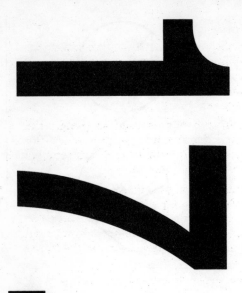

# 17

## probability

**In this chapter you will learn:**

- the meaning of 'probability'
- how to find and use relative frequency
- how to find the probability of single events and of two events
- how to find and use expected frequency.

# 17.1 Introduction

The **probability** of an event is a measure of the likelihood that the event will occur. Probability ranges from 0, when the event is impossible, to 1, when the event is certain to happen, although, as Benjamin Franklin wrote, 'In this world nothing can be said to be certain but death and taxes.'

Sometimes information from the past is used to predict the probabilities of future events. For example, since about 1700, life insurance companies have used mortality tables to estimate life expectancies and thus to set their premiums.

The study of probability has its origins in the mid-sixteenth century, when mathematicians, notably Cardan and Tartaglia, attempted to answer questions arising in gambling. It was to be another 100 years, however, before the Frenchmen Blaise Pascal and Pierre de Fermat developed probability theory, in effect, the laws of chance.

Today, probability is widely used in science and industry, its applications ranging from genetics and quantum mechanics to insurance. Laplace, himself a leading figure in the field of probability, called it a science which began with play but evolved into 'the most important object of human knowledge'.

You can show the probability of an event on the **probability scale** in Figure 17.1.

| Impossible | Unlikely | Evens | Likely | Certain |
|---|---|---|---|---|
| 0 | | 0.5 | | 1 |
| you will get a 7 when you roll a dice | | a coin will come down tails | | the sun will rise tomorrow |

**Figure 17 .1**

# 17.2 Relative frequency

The relative frequency of an event is defined as $\dfrac{\text{number of times event occurs}}{\text{total number of trials}}$. One way of estimating the probability of an event is to carry out an experiment, or trial, a number of times and to find the relative frequency of the event.

For example, if you had a bag of coloured beads and wanted to estimate the probability that you will pick a red bead, you could, without looking, pick a bead from the bag, record its colour and replace it. Suppose that you carried out this experiment 20 times and picked a red bead on 9 occasions. Then

$$\text{relative frequency} = \frac{\text{number of times event occurs}}{\text{total number of trials}} = \frac{9}{20}$$

The relative frequency, $\frac{9}{20}$, estimates the probability of picking a red bead. This result may also be given as a decimal or a percentage but not in any other form. The more times you carry out the experiment, the more accurate your estimate of the probability is likely to be.

### EXERCISE 17.1

1 John spun a coin 50 times. It came down tails 29 times. Find the relative frequency of it coming down tails. Give your answer as
   (a) a fraction, (b) a decimal, (c) a percentage.

2 Kate spun a coin 80 times. It came down heads 34 times. Find the relative frequency of it coming down
   (a) heads, (b) tails.

3 David rolled a dice 100 times. She got a six 22 times. Find the relative frequency of her getting a six.

4 Without looking, Sharon picked a bead from a bag of coloured beads, recorded its colour and then replaced it. She did this 60 times and picked a blue bead on 12 occasions. Find an estimate for the probability that she will pick a blue bead.

5 Paul picked a card from a pack, recorded its suit and then replaced it. He did this 40 times and picked a heart on 12 occasions. Find an estimate for the probability that he will pick a heart.

## 17.3 Probability of a single event

You can find the probability of an event using
$$\text{probability} = \frac{\text{number of successful outcomes}}{\text{total number of possible outcomes}}$$

If you spin a coin, there are two possible outcomes, heads or tails. If the coin is **fair**, or **unbiased**, each of these outcomes is equally likely. Taking a head as the 'successful' outcome,
   probability of a head $=\frac{1}{2}$.

Instead of repeatedly writing the word 'probability', you can use the letter P and write simply, $P(\text{head}) = \frac{1}{2}$.

As probability is expressed as a fraction, decimal or percentage, you could also write P(head) = 0.5 or P(head) = 50%.

---

**Example 17.3.1**

A fair dice is rolled. Find the probability that the number obtained is
(a) a 5,        (b) an odd number,
(c) a 2 or a 3,    (d) less than 7.

There are six possible outcomes and they are all equally likely.

(a) As there is only one 5 on the dice, $P(5) = \frac{1}{6}$.

(b) There are three odd numbers on the dice, so $P(\text{odd no.}) = \frac{3}{6} = \frac{1}{2}$.

(c) A 2 *or* a 3 is a 'success', $P(2 \text{ or } 3) = \frac{2}{6} = \frac{1}{3}$.

(d) As all the numbers on a dice are less than 7, the number obtained will be less than 7.
Thus, $P(\text{number less than } 7) = 1$.

---

In part (c) of Example 17.3.1, $P(2) = \frac{1}{6}$ and $P(3) = \frac{1}{6}$, so, in this case, $P(2 \text{ or } 3) = P(2) + P(3)$, and you could have added the probabilities. This is because the two events cannot both occur and are said to be **mutually exclusive**.

If the events are not mutually exclusive, adding the probabilities would be incorrect.

For example, if you wanted to find the probability of an odd number or a multiple of 3, then the numbers 1, 3, 5 and 6 satisfy the criteria, so

P(odd number or a multiple of 3) $= \frac{4}{6} = \frac{2}{3}$.

But P(odd number) $= \frac{3}{6} = \frac{1}{2}$ and P(multiple of 3) $= \frac{2}{6} = \frac{1}{3}$, so

P(odd number or a multiple of 3)
$\neq$ P(odd number) + P(multiple of 3).

(The sign $\neq$ means 'is not equal to.')

This is because, the number 3 satisfies both criteria, and the events 'odd number' and 'multiple of 3' are not mutually exclusive.

In the following example, the term 'at random' means that every bead has an equal chance of being selected.

### Example 17.3.2

There are 9 beads in a bag. 2 beads are red; 3 beads are green and 4 beads are blue. A bead is picked at random. Find the probability it will be
(a) red,  (b) red or green,  (c) black,  (d) not red.

(a) There are 2 red beads, so $P(\text{red}) = \frac{2}{9}$.
(b) 5 beads are red or green, so $P(\text{red or green}) = \frac{5}{9}$. (Note that these two events are mutually exclusive.)
(c) There are no black beads in the bag so it is impossible for one to be picked. Therefore $P(\text{black}) = 0$.
(d) 7 of the beads are not red, so $P(\text{not red}) = \frac{7}{9}$.

In part (d), $P(\text{not red}) = 1 - P(\text{red})$. In general, if the probability that an event will occur is $p$, then the probability that the event will not occur is $1 - p$. For example, if the probability that it will rain tomorrow is 0.7, the probability that it will not rain tomorrow is $1 - 0.7 = 0.3$.

### EXERCISE 17.2

1 A fair dice is rolled. Find the probability of obtaining
   (a) a 6,              (b) an even number,
   (c) a prime number,  (d) a 4 or a 5.

2 A fair spinner has the numbers 1, 2, 3, 4 and 5 on it. Find the probability that it will land on
   (a) a 1,   (b) an odd number,   (c) a 3 or a 4,
   (d) a 6,   (e) an even number or a prime number,
   (f) a number other than 5.

3 There are 10 beads in a bag. 5 beads are black; 3 beads are white and 2 beads are red. A bead is picked at random. Find the probability that it will be
   (a) white,        (b) black or red,
   (c) not white,    (d) not blue.

4 A card is picked at random from a pack of 52. Find the probability that it will be
   (a) the queen of hearts,  (b) a king,
   (c) a club,               (d) a red card,
   (e) a king or a queen,    (f) an ace or a spade.

5  A letter is chosen at random from the word MATHEMATICS. Find the probability that it will be
   (a) an H,      (b) an M,        (c) a B,
   (d) a vowel,   (e) an S or a T,  (f) other than an A.

6  In a raffle, 500 tickets are sold. George buys one ticket and Julia buys five. There is one winning ticket. Find the probability that
   (a) George will win,    (b) Julia will win.

7  The probability that Sally will be late for work tomorrow is $\frac{2}{9}$. What is the probability that she will not be late for work?

8  My drawer contains grey socks, brown socks and black socks. I pick a sock at random. The probability that it will be a grey sock is 0.5 and the probability that it will be a brown sock is 0.2. What is the probability that it will be a black sock?

# 17.4  Two events

If you spin two coins, you can get two heads, two tails or one of each. If you spin two coins 100 times, you will find that you get two heads about 25 times, two tails about 25 times and one of each about 50 times. To explain this, there are four possible outcomes which you can show in a list.

   HH   HT   TH   TT

Alternatively, you can show the possible outcomes on a diagram like Figure 17.2; each cross represents one of the four possible outcomes.

When you spin two coins, the probabilities that you will get two heads, two tails and a head and a tail are respectively $\frac{1}{4}, \frac{1}{4}$ and $\frac{2}{4} = \frac{1}{2}$. The sum of these three probabilities is 1, as one of them is certain to happen.

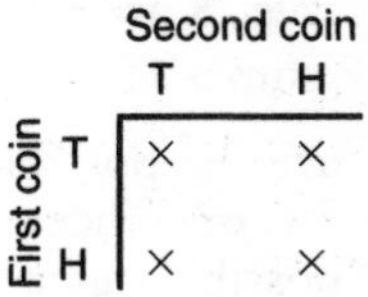

**Figure 17.2**

When you spin two coins, the outcome for one coin has no effect on the outcome for the other coin. The events are said to be **independent**.

Rolling two dice is another example of two independent events. You can show the 36 possible outcomes as a list, rather like coordinates.

$$(1,1) \quad (2,1) \quad (3,1) \quad (4,1) \quad (5,1) \quad (6,1)$$
$$(1,2) \quad (2,2) \quad (3,2) \quad (4,2) \quad (5,2) \quad (6,2)$$
$$(1,3) \quad (2,3) \quad (3,3) \quad (4,3) \quad (5,3) \quad (6,3)$$
$$(1,4) \quad (2,4) \quad (3,4) \quad (4,4) \quad (5,4) \quad (6,4)$$
$$(1,5) \quad (2,5) \quad (3,5) \quad (4,5) \quad (5,5) \quad (6,5)$$
$$(1,6) \quad (2,6) \quad (3,6) \quad (4,6) \quad (5,6) \quad (6,6)$$

You can also show the outcomes on a diagram, where each cross represents one of the 36 possible outcomes. See Figure 17.3. Using either the list or the figure, you can find, for example, that the probability of a double six is $\frac{1}{36}$ or the probability that the sum of the numbers on the two dice is 4 is $\frac{3}{36}$ or $\frac{1}{12}$. In the latter case the table in Figure 17.4 showing the sum of the two numbers for all 36 possible outcomes is helpful.

Second dice

|   | 1 | 2 | 3 | 4 | 5 | 6 |
|---|---|---|---|---|---|---|
| 1 | × | × | × | × | × | × |
| 2 | × | × | × | × | × | × |
| 3 | × | × | × | × | × | × |
| 4 | × | × | × | × | × | × |
| 5 | × | × | × | × | × | × |
| 6 | × | × | × | × | × | × |

First dice

**Figure 17.3**

You can use the table to help you find, for example, the probability that the sum of the numbers on the two dice is more than 10. Of the 36 sums in the table, 3 are more than 10, so $P(\text{sum} > 10) = \frac{3}{36} = \frac{1}{12}$.

Second dice

| + | 1 | 2 | 3 | 4 | 5 | 6 |
|---|---|---|---|---|---|---|
| 1 | 2 | 3 | 4 | 5 | 6 | 7 |
| 2 | 3 | 4 | 5 | 6 | 7 | 8 |
| 3 | 4 | 5 | 6 | 7 | 8 | 9 |
| 4 | 5 | 6 | 7 | 8 | 9 | 10 |
| 5 | 6 | 7 | 8 | 9 | 10 | 11 |
| 6 | 7 | 8 | 9 | 10 | 11 | 12 |

First dice

**Figure 17.4**

There are 6 possible outcomes for one dice and 36 ($6 \times 6$) possible outcomes for two dice. Similarly, there are 2 possible outcomes for one coin and 4 ($2 \times 2$) possible outcomes for two coins. This suggests that you multiply the number of possible outcomes for each of the separate items to find the number of possible outcomes for a combination of the two items.

**Example 17.4.1**

A coin is spun and a dice is rolled.
(a) How many possible outcomes are there?
(b) Show all the possible outcomes on a diagram.
(c) Find the probability of obtaining (i) a head and a 5,
(ii) a tail and an even number.

(a) There are two possible outcomes for a coin and six
possible outcomes for a dice. So, for the coin and the
dice together, there are $2 \times 6 = 12$ possible outcomes.
(b) See the figure.
(c) (i) $P(H,5) = P(H \text{ and } 5) = \frac{1}{12}$.
(ii) Three possible outcomes have a tail and an even
number: they are (T,2), (T,4) and (T,6).
So P (tail and even number) $= \frac{3}{12} = \frac{1}{4}$.

Dice

|  |  | 1 | 2 | 3 | 4 | 5 | 6 |
|---|---|---|---|---|---|---|---|
| Coin | H | × | × | × | × | × | × |
|  | T | × | × | × | × | × | × |

## EXERCISE 17.3

1 Two dice are rolled. Find the probability that
(a) the numbers on the dice are the same,
(b) there is a 6 on at least one of the dice,
(c) the sum of the numbers is less than 5.

2 Two dice are rolled. The sum of the numbers on the dice is
found.

(a) Complete the table to show the probability of obtaining
each of the possible sums.

| Sum | 2 | 3 | 4 | 5 | 6 | 7 | 8 | 9 | 10 | 11 | 12 |
|---|---|---|---|---|---|---|---|---|---|---|---|
| Probability |  |  |  |  |  | $\frac{5}{36}$ |  |  |  |  | $\frac{1}{36}$ |

(b) What is the most likely sum?
(c) Find the sum of the probabilities. Explain your answer.

3 A coin is spun and a dice rolled. Find the probability of
obtaining
(a) a tail and a 5,
(b) a head and 2 or a head and 3,
(c) a head and a prime number,
(d) a tail and a number less than 4.

4 Two dice are rolled. The difference between the numbers on the dice is found.
   (a) Show all the possible outcomes in a table.
   (b) Find the probability that the difference will be (i) 2, (ii) 5.
   (c) What is the most likely difference?

5 A spinner has the numbers 1, 2 and 3 on it. A coin and this spinner are spun.
   (a) How many possible outcomes are there?
   (b) List all the possible outcomes.
   (c) Draw a diagram to show all the possible outcomes.
   (d) Find the probability of obtaining
      (i) a tail and a 3,   (ii) a head and an odd number,
      (iii) a tail and a factor of 6.

6 Two spinners are spun. Each has the numbers 1, 2, 3, 4 and 5 on it. The sum of the two numbers on the spinners is found.
   (a) How many possible outcomes are there?
   (b) Show all the possible outcomes in a table.
   (c) Find the probability that the sum will be (i) 5, (ii) 3 or 4.
   (d) What is the most likely sum?

7 Out of three beads in a bag, 1 bead is red, 1 is green and 1 is blue. A dice is thrown and a bead is picked at random from the bag.
   (a) How many possible outcomes are there?
   (b) Find the probability of obtaining
      (i) a red bead and a 4,   (ii) a blue bead and a 5 or a 6,
      (iii) a green bead and a number greater than 4.

8 One spinner has the numbers 1, 2, 3 and 4 on it. Another spinner has the numbers 2, 3 and 5 on it. The two spinners are spun. Find the most likely sum of the two numbers.

## 17.5 Tree diagrams

Another way of showing the possible outcomes when a coin is spun is to use a tree diagram (see Figure 17.5). The probability of each outcome is written on the branch leading to it. The sum of the probabilities on the branches is 1, because you are certain to get one of the outcomes. You can extend the diagram to show the possible outcomes when two coins are spun. (See Figure 17.6.)

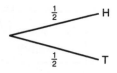

**Figure 17.5**

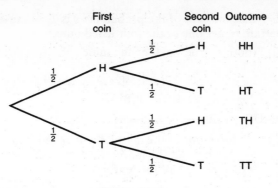

**Figure 17 .6**

At the start of Section 17.4, you saw that the probability of getting two heads is $\frac{1}{4}$. You can find this probability by working out $\frac{1}{2} \times \frac{1}{2}$, *multiplying* the probabilities on the two branches leading to the outcome HH. This only works for independent events.

In general, you can find the probability that two independent events both occur by multiplying their probabilities.

---

**Example 17.5.1**

There are 5 marbles in a bag, 2 of the marbles are red and the other 3 marbles are blue. A marble is picked at random. Its colour is noted and the marble is replaced in the bag. A marble is again picked at random and its colour noted.
(a) Draw a tree diagram showing all the outcomes and the probability of each one.
(b) Find the probability that (i) both marbles will be the same colour, (ii) at least one of the marbles will be red.

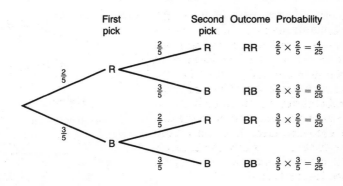

(Notice that the sum of the probabilities is 1, because you are certain to get one of the four listed outcomes.)

(b) (i) P(both marbles will be the same colour) = P(RR or BB).

As the outcomes RR and BB are mutually exclusive,

P(RR or BB) = P(RR) + P(BB), so

$P(RR) + P(BB) = \frac{4}{25} + \frac{9}{25} = \frac{13}{25}$.

(ii) 'At least' one of the marbles will be red means that one or both of the marbles will be red, so
P(at least one of the marbles will be red)

$$= P(RR \text{ or } RB \text{ or } BR)$$
$$= P(RR) + P(RB) + P(BR)$$
$$= \frac{4}{25} + \frac{6}{25} + \frac{6}{25} = \frac{16}{25}.$$

Alternatively, P(at least one of the marbles will be red)

$$= 1 - P(BB)$$
$$= 1 - \frac{9}{25} = \frac{16}{25}.$$

It is not always necessary to draw a complete tree diagram. Sometimes not all the branches are relevant to the problem.

---

### Example 17.5.2

I have to go through two sets of traffic lights on my way to work. The probability that the first set will be on green is 0.7 and the probability that the second set will be on green is 0.8. Find the probability that the first set will be on green and the second set will not be on green.

P(second set will not be on green) = 1 − 0.8 = 0.2.

The relevant part of the tree diagram is shown in the diagram.

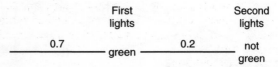

The probability that the first set will be on green and the second set will not be on green = 0.7 × 0.2 = 0.14.

**1** There are 8 marbles in a bag, 5 of the marbles are red and the other 3 marbles are green. A marble is picked at random. Its colour is noted and the marble is replaced in the bag. A marble is again picked at random and its colour noted.
  (a) Draw a tree diagram showing all the outcomes and the probability of each one.
  (b) Find the probability that
      (i) the marbles will be different colours,
      (ii) neither marble will be green.

**2** Sharon has a biased coin. The probability that it will come down heads is $\frac{2}{3}$. Sharon spins the coin twice.
  (a) Draw a tree diagram showing all the outcomes and the probability of each one.
  (b) Find the probability that she will get
      (i) at least one head, (ii) either two heads or two tails.

**3** The probability that Tom will pass a science test is 0.7 and the probability that he will pass a French test is 0.4. Find the probability that he will
  (a) pass both tests,                    (b) fail both tests,
  (c) pass at least one of the tests.

**4** A card is picked at random from a pack. It is replaced and a card is again picked at random. Find the probability of picking
  (a) two red cards,
  (b) two aces,
  (c) first a heart and second a king,
  (d) a heart and a king (in either order),
  (e) at least one diamond.

**5** The letters in the word LONDON are put in one bag and the letters in the word LIVERPOOL are put in another bag. A letter is picked at random from each bag. Find the probability of picking
  (a) two Os,      (b) two vowels,      (c) at least one L.

**6** There are 9 beads in a bag, 2 of the beads are red, 3 are blue and the rest are black. A bead is picked at random and then replaced. A bead is again picked at random. Find the probability that
  (a) both beads will be red,
  (b) both beads will be the same colour,
  (c) exactly one of the beads will be black.

**7** Holly catches a bus to work every day. The bus may be on time or early or late. The probability that it will be on time is 0.5 and the probability that it will be early is 0.3. Find the

probability that, in the next two days, the bus will be
(a) early twice, (b) late twice, (c) on time at least once.

8 The probability that Melchester Rovers will win a soccer match is 0.6 and the probability that they will draw is 0.3. Find the probability that they will
(a) win the next two matches,
(b) lose the next two matches,
(c) draw at least one of the next two matches.

## 17.6  Expected frequency

The laws of chance are of most use when applied to a large number of trials rather than to a particular trial. For example, if you spin a coin 100 times, you would expect to get heads about 50 times. This is the **expected frequency**. It may be calculated by finding the product of the probability, $\frac{1}{2}$, and the number of trials, 100.

**Expected frequency = probability × number of trials.**

Similarly, it is not the fact that the probability that a 30-year-old man will live to the age of 70 is 0.4514 which is of interest to a life assurance company. What this means to them is that about 4500 of every 10 000 30-year-old men will live to the age of 70. This aspect of probability is known as the law of large numbers.

---

**Example 17.6.1**

A die is rolled 600 times. Find the expected frequency of a 5.

P(five) $= \frac{1}{6}$. The number of trials is 600. Using the formula

expected frequency = probability × number of trials,

expected frequency $= \frac{1}{6} \times 600 = 100$.

---

**Example 17.6.2**

Two dice are rolled 180 times. How many times would you expect their sum to be   (a) 2, (b) 11 or 12?

(a)  P(sum is 2) $= \frac{1}{36}$. The number of trials is 180, so
expected frequency $= \frac{1}{36} \times 180 = 5$.

(b)  P(sum is 11 or 12) $= \frac{3}{36} = \frac{1}{12}$.

Expected frequency $= \frac{1}{12} \times 180 = 15$.

1 A coin is spun 500 times. How many times would you expect it to come down heads?

2 A dice is rolled 300 times. How many times would you expect
(a) a 4,      (b) a 2 or a 3,      (c) an even number.

3 There are 8 counters in a bag, 5 of the counters are green; 2 are red and 1 is black. A counter is picked at random from the bag and then replaced. In 200 picks, how many times would you expect to get
(a) a green counter,
(b) a red counter or a black counter,
(c) a counter which is not black.

4 The probability that a baby will be a boy is 0.52. How many boys would you expect in 1000 births?

5 The probability that it will rain on any day in a month is 40%. How many rainy days would you expect in a month of 30 days?

6 Two coins are spun 200 times. How many times would you expect to get
(a) two tails, (b) a head and a tail, (c) at least one head?

7 Two dice are rolled 720 times. How many times would you expect their sum to be
(a) 4,      (b) more than 9,      (c) 7 or 8.

8 A coin is spun and a dice is rolled. This is done 300 times. How many times would you expect to get
(a) a tail and a 4,      (b) a head and an odd number?

9 A card is picked at random from a pack and a coin is spun. This is done 260 times. How many times would you expect an ace and a head?

10 A card is picked at random from a pack and a dice is spun. This is done 400 times. How many times would you expect a diamond and a prime number?

# 8

## three-dimensional shapes

**In this chapter you will learn:**

- the names of solids
- how to find the surface area of a solid
- how to find and use the volume of a cuboid
- how to find the volume of a prism.

# 18.1 Introduction

Three-dimensional shapes, or solids, have length, breadth and height (or thickness). In other words, they take up space in all directions. An example is the **cube**, shown in Figure 18.1.

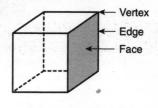

**Figure 18.1**

One **face** has been shaded. A cube has 6 faces altogether, all of them squares. The line where two faces meet is called an **edge**; the cube has 12 edges. Edges meet at a **vertex**, or corner; the cube has 8 vertices.

The **cuboid**, or rectangular box, shown in Figure 18.2 has the same numbers of faces, edges and vertices as the cube but its faces are rectangular, rather than square.

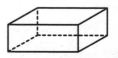

**Figure 18.2**

There are several categories of three-dimensional shapes. A **prism** is a solid with a constant cross section. That is, if you cut through a prism parallel to one of its ends, you get the same shape all along its length. This shape is called the **cross section** of the prism. Both a cube and a cuboid are examples of prisms. A prism is usually defined in terms of the shape of its cross section. However, a circular prism has a special name and is called a **cylinder**. Three prisms are shown in Figure 18.3.

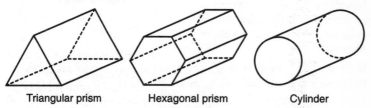

Triangular prism          Hexagonal prism          Cylinder

**Figure 18.3**

The term **pyramid** probably evokes thoughts of the square-based pyramids built over 4000 years ago by the Ancient Egyptians. In fact, the base may be any shape but, whatever shape the base is, lines drawn from the edge of the base must meet at a point, called the **apex**. If the apex is vertically above the centre of the base, the pyramid is called a **right** pyramid.

A pyramid is usually defined in terms of the shape of its base. This is often a polygon, in which case the other faces are triangles. Two pyramids have special names. A pyramid with a triangular base is called a **tetrahedron**; if each face is an equilateral triangle, it is a regular tetrahedron. A pyramid with a circular base is called a **cone**. Three pyramids are shown in Figure 18.4.

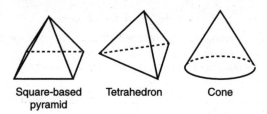

Square-based    Tetrahedron    Cone
pyramid

**Figure 18.4**

Another common three-dimensional shape is a **sphere**, which, like a ball, is circular in every direction. (See Figure 18.5.)

If every face of a solid is a polygon, the solid is called a **polyhedron**. Of the solids illustrated so far, a cube, a cuboid, a triangular prism, a hexagonal prism, a square-based pyramid and a tetrahedron are polyhedra but a cylinder, a cone and a sphere are not.

Sphere

**Figure 18.5**

If all the edges of a solid are the same length and all its faces are the same shape and all its angles are the same, it is called a **regular solid**. The cube and the regular tetrahedron are the only regular solids illustrated so far but there are five regular solids altogether. Often called **Platonic solids,** they have fascinated mathematicians for 3000 years.

There is an interesting relationship between the number of faces ($F$), the number of vertices ($V$) and the number of edges ($E$) of a polyhedron.

Expressed as a formula, it is $F + V = E + 2$, a result often known as Euler's theorem. It may have been known to Archimedes (c 225 BC), but Descartes, in about 1635, was the first to state it in this form. The Swiss mathematician Leonhard Euler proved the result in 1752.

### EXERCISE 18.1

1  How many faces, vertices and edges do each of these of these polyhedra have?
   (a) triangular prism,          (b) hexagonal prism,
   (c) square-based pyramid,   (d) tetrahedron.

2  (a) Draw a sketch of a prism which has a regular pentagon as its cross section.
   (b) How many faces, vertices and edges does the prism have?

3  A polyhedron has 10 faces and 16 vertices. Work out how many edges it has.

4  Find out the names of the other three Platonic solids and the number of faces each of them has.

## 18.2  Nets and surface area

A **net** of a solid shape is a flat shape which can be cut out and folded to make the solid shape. Figure 18.6 shows a sketch of a cuboid and one of its possible nets.

This is not the only possible net of the cuboid; there are several others. Nets are helpful in finding the surface area of a three-dimensional shape. For example, to find the surface area of the above cuboid, you can find the total area of the six rectangles which make up its net. You can shorten the working by using the fact that there are three pairs of congruent rectangles in the net.

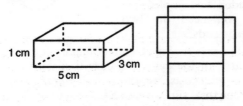

**Figure 18.6**

The surface area in cm$^2$ $= 2 \times (5 \times 3) + 2 \times (5 \times 1) + 2 \times (3 \times 1)$
$= 46 \, \text{cm}^2.$

In general, the surface area, $A$, of a cuboid with length $l$, breadth $b$ and height $h$ is given, in appropriate units, by the formula

$$A = 2lb + 2hl + 2bh.$$

---

### Example 18.2.1

The figure shows a solid cylinder. The radius of the circular end of the cylinder is $r$ and the height of the cylinder is $h$.

(a) Draw a net of the cylinder.

(b) Find formulae for (i) the area of the curved surface ($A$) and (ii) the total surface area ($S$) of the cylinder.

(c) Use your formula for $S$ to find the total surface area of a cylinder with an end radius of 3.6 cm and a height of 8.3 cm. Give your answer correct to three significant figures.

(a) The net of the curved surface of a cylinder is a rectangle, the shape that you get if you remove the label from a tin of soup and flatten it. The length of the rectangle is equal to the circumference of the circular end of the cylinder. To complete the net you need the two circular ends.

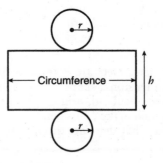

(b) (i) The area $A$ of the rectangle (circumference $\times h$) gives the area of the cylinder's curved surface. Writing the circumference as $2\pi r$, the area $A = 2\pi rh$.

(ii) The total surface area $S$ is the sum of the area of the curved surface and the areas of the two circular ends. Thus $S = 2\pi rh + 2\pi r^2$. You can factorize this expression if you wish, to get $S = 2\pi r(h + r)$.

(c) To find the total surface area, you substitute 3.6 for
$r$ and 8.3 for $h$ in one of the formulae for $S$. Using the
factorized form,

$$S = 2\pi r(h + r) = 2 \times \pi \times 3.6 \times (8.3 + 3.6)$$
$$= 2 \times \pi \times 3.6 \times 11.9 = 269, \text{correct to}$$

three significant figures.

The total surface area is $269\,\text{cm}^2$.

## EXERCISE 18.2

1 The sides of a cube are 7 cm long.
(a) Draw a sketch of a net of the cube.
(b) Find the surface area of the cube.

2 The length of a cuboid is 10 cm. Its breadth is 6 cm and its
height is 4 cm. Find its surface area.

3 The figure shows a triangular prism. The cross section
of the prism is a right-
angled triangle with sides
3 cm, 4 cm and 5 cm long.
The length of the prism is
8 cm.
(a) Draw a sketch of the net of
the prism.
(b) Find the surface area of the prism.

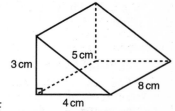

4 A pyramid has a square base with sides 8 cm long. Its
other faces are isosceles triangles with a vertical height
of 10 cm.
(a) Draw a sketch of the net of the pyramid.
(b) Find the surface area of the pyramid.

5 The radius of the ends of a solid cylinder is 4.7 cm and its
length is 7.2 cm. Find its total surface area.

6 A hollow cylinder is open at one end. Its radius is 5.3 cm and
its length is 9.1 cm.
(a) Draw a net of the cylinder.
(b) Find the total surface area of the cylinder.

## 18.3 Volume of a cuboid

The volume of a three-dimensional shape is a measure of the amount of space it occupies.

Figure 18.7 shows a cube with edges 1 cm long. It is called a centimetre cube and its volume is 1 cubic centimetre (written $1 \text{ cm}^3$).

**Figure 18.7**

The volume of the cuboid in Figure 18.8 is the number of centimetre cubes it contains.

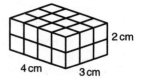

**Figure 18.8**

There are $4 \times 3 = 12$ centimetre cubes in each layer and there are two layers. So the number of centimetre cubes is $4 \times 3 \times 2 = 24$. The volume of the cuboid is $24 \text{ cm}^3$.

In general,

**volume of a cuboid = length × breadth × height.**

The volume, $V$, of a cuboid with length $l$, breadth $b$ and height $h$ is given by the formula $V = lbh$.

For the volumes of larger cuboids, you could use cubic metres, written $\text{m}^3$, and, for smaller cuboids, cubic millimetres written $\text{mm}^3$. You must ensure, however, that you use the same units for the length, the breadth and the height.

---

**Example 18.3.1**

The length of a cuboid is 5 m; its breadth is 3 m and its height is 40 cm. Find its volume in $\text{m}^3$.

The units of the measurements must be the same, so the height is 0.4 m. If $V\text{m}^3$ is the volume, $V = 5 \times 3 \times 0.4 = 6$; the volume is therefore $6 \text{ m}^3$.

---

Sometimes, the volume of a cuboid will be given and you will have to find one of its dimensions.

**Example 18.3.2**

The volume of a cuboid is $336\,\text{cm}^3$. Its length is 8 cm and its breadth is 7 cm. Find its height.

Let the height of the cuboid be $h$ cm. Then $336 = 8 \times 7 \times h$, giving $56h = 336$ and $h = \dfrac{336}{56} = 6$. The height of the cuboid $= 6$ cm.

## EXERCISE 18.3

1 The edges of a cube are 9 cm long. Find its volume.

2 Find the volume of cuboids with these dimensions.
   (a) Length 8 cm, breadth 5 cm and height 4 cm,
   (b) Length 10 m, breadth 7 m and height 3 m.

3 The length of a cuboid is 2 m; its breadth is 30 cm and its height is 20 cm. Find its volume in $\text{cm}^3$.

4 Find the number of cubic metres of concrete needed for a path 12 m long, 1.5 m wide and 10 cm thick.

5 The volume of a cuboid is $60\,\text{cm}^3$. Its length is 5 cm and its breadth is 4 cm. Find its height.

6 The volume of a cuboid is $400\,\text{cm}^3$. Its length is 10 cm and its height is 8 cm. Find its breadth.

7 A water tank measures 1.4 m by 80 cm by 60 cm. Find its capacity (the volume of water it can hold) in litres.

8 A room measures 10 m by 6 m by 3 m. The volume of airspace needed for one person is $5\,\text{m}^3$. Find the maximum number of people who may use the room at the same time.

9 How many cubic centimetres are there in a cubic metre?

10 How many packets of tea measuring 12 cm by 4 cm by 4 cm may be packed in a box measuring 48 cm by 24 cm by 16 cm?

# 18.4 Volume of a prism

Figure 18.9 shows a prism with a right-angled triangle as its cross section.

The volume of the prism will be half the volume of a cuboid measuring 4 cm by 5 cm by 9 cm.

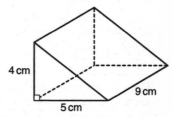

**Figure 18.9**

If $V$ cm$^3$ is the volume, then $V = \frac{1}{2} \times 5 \times 4 \times 9 = 90$ cm$^3$.

$\frac{1}{2} \times 5 \times 4$ cm$^2$ is the area of the end of the prism, that is, its area of cross section. This demonstrates the result

> **volume of a prism = area of cross section × length.**

This is sometimes written as

> **volume of a prism = end area × length.**

---

### Example 18.4.1

The figure shows a prism with a trapezium as its cross section. Find the volume of the prism.

Let area of cross section be $A$ cm$^2$.

Then $A = \frac{1}{2} \times$ (sum of parallel sides) × (distance between them)

$$= \frac{1}{2} \times (8 + 6) \times 4 = 28.$$

If $V$ cm$^3$ is the volume, $V = 28 \times 10 = 280$.

The volume is $280$ cm$^3$.

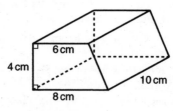

---

A cylinder with radius $r$ and height $h$ has an area of cross section of $\pi r^2$. So you can find the volume, $V$, of a cylinder using this formula.

$$V = \pi r^2 \times h = \pi r^2 h.$$

---

### Example 18.4.2

The radius of the end of a cylinder is 3.8 cm and its height is 8.4 cm. Find the volume of the cylinder, correct to three significant figures.

Volume $= ((\pi \times 3.8^2) \times 8.4)$ cm$^3$ $= 381.06$ cm$^3$.

The volume is $381$ cm$^3$, correct to three significant figures.

# EXERCISE 18.4

1 Find the volume of the triangular prism shown in the left-hand figure below.

2 The right-hand figure below shows a prism with a trapezium as its cross section. Find its volume.

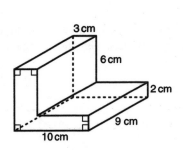
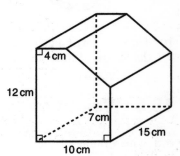

3 The radius of the end of a cylinder is 5.7 cm. Its height is 13.6 cm. Find the volume of the cylinder correct to three significant figures.

4 Find the volume of each of the prisms shown in the figure.

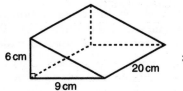
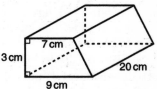

5 A prism has a volume of 120 cm$^3$. Its area of cross section is 15 cm$^2$. Find the length of the prism.

6 The left-hand figure overleaf shows the cross section of a swimming pool which is 25 m long. Find the volume of water in the pool when it is full.

7 The right-hand figure shows the cross section of a barn which is 12 m long. Find the volume of the barn.

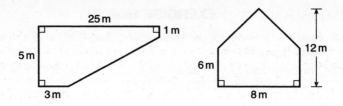

8 The left-hand figure below shows the cross section of a circular pipe. The internal and external diameters are 10 cm and 12 cm. Find the volume of material used to make a 2-metre length of pipe.

9 The right-hand figure below shows the cross section of a prism. It is a regular hexagon. The prism is 20 cm long. Find its volume.

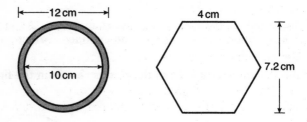

10 The volume of a cylinder is 300 cm³. Its radius is 5 cm. Find the length of the cylinder correct to three significant figures.

## 18.5 Weight of a prism

Figure 18.10 shows a steel rod. The volume of the rod is 200 cm³. You can find the **weight** of the rod if you know the weight of 1 cubic centimetre of steel. In fact, steel weighs 7.7 grams per cm³ (the **density** of steel).

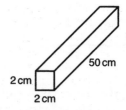

**Figure 18.10**

You can then find the weight of the rod by multiplying the volume of the rod by its density.

Its weight in grams is $200 \times 7.7 = 1540$.

The weight is 1540 g or 1.54 kg.

There is a problem with the use of the word 'weight'. Strictly speaking, what has been found is the **mass** of the rod, and this is the term which is used in science. However, in everyday language people use the word 'weight' rather than 'mass', so, even though it is incorrect, 'weight' is the word which will be used in this book.

In general,
  **weight = volume × density**.

Density may also be measured in other units, for example, kilograms per m³.

## EXERCISE 18.5

1  The volume of a brick is 1400 cm³. The density of the brick is 1.6 grams per cm³. Find the weight of the brick.

2  A concrete block in the shape of a cuboid measures 10 cm by 21.5 cm by 44 cm. The density of concrete is 2.3 grams per cm³. Find the weight of the block.

3  A solid cylinder of gold has a radius of 2 cm and a height of 5 cm. The density of gold is 19.3 grams per cm³. Find the weight of the cylinder correct to three significant figures.

4  The figure shows the cross section of a steel girder. The density of the steel in the girder is 7700 kilograms per m³. The length of the girder is 5 m. Find its weight.

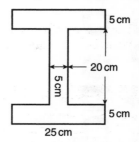

5  Each side of a square brass plate is 4 cm long. The plate is 5 mm thick. A circular hole with a diameter of 2 cm is drilled in the plate. The density of brass is 8.3 grams per cm³. Find the weight of brass which remains. Give your answer correct to the nearest gram.

# ratio and proportion

**In this chapter you will learn:**

- the meaning of 'ratio'
- how to simplify ratios
- how to use scales
- how to use ratio
- how to use direct proportion and inverse proportion.

# 19.1 What is a ratio?

1:72 is the scale of a model aeroplane. 1:10 appears on a road sign. 4:3:2 is printed on a sack of fertilizer. Each of these is a **ratio**.

You use a ratio to compare quantities. For example, if there are 9 men and 7 women in a room, the ratio of men to women is 9:7. The order of the numbers is important, as 7:9 is the ratio of women to men. You may also write these ratios as 9 to 7 and 7 to 9 respectively but, in this chapter, colons will be used.

The numbers 9 and 7 have no common factors and so you cannot simplify the ratio 9:7. If, however, the numbers in a ratio do have a common factor, you can simplify the ratio and find its simplest form using a method similar to the one you used to find the simplest form of a fraction in Chapter 03, Section 3.4.

To find the simplest form of the ratio 24:30, for example, you divide both numbers by 6, the highest common factor of 24 and 30. Thus, the simplest form of 24:30 is 4:5.

If the quantities in the ratio have units and the units are the same, such as 25 cm:15 cm, you remove the units, obtaining 25:15 in this case.

Dividing both numbers by 5 gives its simplest form as 5:3.

If the units used for the quantities in the ratio are not the same, you must start by expressing both quantities in the same units.

---

**Example 19.1.1**

Find the simplest form of the ratio 27 mins:$1\frac{1}{2}$ hours.

| | |
|---|---|
| Change $1\frac{1}{2}$ hours to minutes. | 27 mins:90 mins |
| Remove the units. | 27:90 |
| Divide both numbers by 9. | 3:10 is the simplest form. |

---

To compare ratios, it is sometimes useful to express them in the form 1:$n$ or $n$:1. For example, if one school has 808 students and 47 teachers, while another has 1025 students and 61 teachers, it is not obvious which school has the smaller ratio of students to teachers. For the first school, the ratio of students to teachers is 808:47. If you divide both numbers by 47, you obtain, 17.2:1, correct to one decimal place. For the second school, the ratio of students to teachers is 1025:61. If you divide both numbers by 61, you obtain 16.8:1, correct to one decimal place. So the second school has the smaller ratio of students to teachers.

### EXERCISE 19.1

1 Find the simplest form of each of these ratios.
(a) 8:12  (b) 6:3  (c) 20:15  (d) 18:30

2 Find the simplest form of each of these ratios.
(a) 24 cm:18 cm          (b) 9 days:27 days
(c) 60 cm:1.5 m          (d) $1\frac{1}{4}$ h:30 min
(e) 40 s:3 min           (f) 32 mm:8 cm
(g) 1.4 kg:420 g         (h) 720 ml:1.2 l

3 The lengths of the sides of two squares are 4 cm and 6 cm. Find, in its simplest form, the ratio of
(a) their perimeters,       (b) their areas.

4 The lengths of the edges of two cubes are 2 cm and 4 cm. Find, in the form 1:$n$, the ratio of
(a) the total lengths of their edges,
(b) their surface areas,                    (c) their volumes.

5 Express these ratios in the form 1:$n$.
(a) 3:12  (b) 2:3  (c) 4:9  (d) 10:8

6 Express these ratios in the form $n$:1.
(a) 18:3  (b) 7:2  (c) 23:10  (d) 12:16

7 Sinton School has 723 students and 45 teachers. Tanville School has 936 students and 57 teachers. For each school, calculate, correct to one decimal place, the ratio of students to teachers in the form $n$:1. Which school has the smaller ratio of students to teachers?

## 19.2 Scales

If a model of a car is made to a scale of 1:50, the model is $\frac{1}{50}$ of the size of the real car, that is, you divide the real car's measurements by 50 to obtain those of the model car. Thus, if the real length of the car is 4.5 m, the length of the model car is 4.5 m ÷ 50 = 0.09 m or 9 cm.

To find the real car's measurements, you multiply the model car's measurements by 50. So, if the height of the model car is 2.8 cm, the real height of the car is 50 × 2.8 cm = 140 cm or 1.4 m.

The scales of maps are sometimes expressed as ratios. The scale tells you the relationship between distances on the map and real distances.

**Example 19.2.1**

The scale of a map is 1:20 000.
(a) Find the real distance represented by a distance of 6 cm on the map.
(b) What distance on the map represents a real distance of 5 km?

(a) 1 cm represents 20 000 cm.
So 6 cm represents $6 \times 20\,000$ cm $= 120\,000$ cm.
Convert 120 000 cm to km by dividing by 100 000.
Thus, 120 000 cm $= 1.2$ km.
6 cm on the map represents a real distance of 1.2 km.

(b) 5 km $= 500\,000$ cm.
Distance on the map $= \dfrac{500\,000}{20\,000}$ cm $= 25$ cm.
25 cm on the map represents a real distance of 5 km.

If you know a measurement on a model or a map and the real measurement it represents, you can find the scale being used. For example, if a length of 4 cm on a model represents a real length of 1.44 m, the ratio of the lengths is 4 cm:1.44 m. Converting 1.44 m to centimetres, the ratio becomes 4 cm:144 cm. You can then remove the units and divide both numbers by 4 to obtain 1:36 as the scale.

**Example 19.2.2**

The distance between two towns is 4.5 km. On a map, the distance between the towns is 9 cm. Find the scale of the map.

The ratio of the distances is 9 cm: 4.5 km.
Convert 4.5 km to cm, that is, by multiplying 4.5 by 100 000.
Ratio is 9 cm: 450 000 cm.
Remove the units and divide both sides by 9.
The scale is 1:50 000.

## EXERCISE 19.2

1 The scale of a model car is 1:100.
(a) The length of the model car is 3.7 cm. Find, in metres, the real length of the car.
(b) The real width of the car is 1.61 m. Find, in centimetres, the width of the model car.

2 A plan of a house is drawn to a scale of 1:50.
  (a) On the plan, the dining room is 14 cm long and 12.4 cm wide. Find, in metres, the real length and width of the dining room.
  (b) The lounge is 8 m long and 6.8 m wide. Find, in centimetres, its length and its width on the plan.

3 The scale of a model aeroplane is 1:72.
  (a) The length of the model aeroplane is 86 cm. Find, in metres, the real length of the aeroplane.
  (b) The wingspan of the aeroplane is 25.2 m. Find, in centimetres, the wingspan of the model.

4 The scale of a map is 1:40 000.
  (a) Find, in kilometres, the real distance represented by a distance of 8 cm on the map.
  (b) What distance on the map represents a real distance of 8 km?

5 The distance between two towns is 7 km. On a map, the distance between the towns is 28 cm. Find, as a ratio, the scale of the map.

## 19.3  Using ratio

You can use ratio to solve a wide variety of problems, many of which fall into one of two categories. The first of these is when you are told the ratio of two quantities and the amount of one of the quantities and have to find the amount of remaining quantity.

For example, a type of mortar is made by mixing sand and cement in the ratio, by weight, of 4:1. This means the weight of sand is four times that of cement. So, if you have 24 kg of sand the weight of cement you have to mix with it is 24 kg divided by 4, that is, 6 kg.

---

**Example 19.3.1**

Brass is an alloy (mixture) containing copper and zinc in the ratio 5:2.

(a) What weight of copper is mixed with 26 g of zinc?
(b) What weight of brass contains 40 g of copper?

(a) The ratio is equivalent to $\frac{5}{2}$:1, so there are $\frac{5}{2}$ grams of copper to every gram of zinc.
   Weight of copper is $\frac{5}{2} \times 26$ g $= 65$ g.

(b) The ratio is equivalent to $1:\frac{2}{5}$, so there are $\frac{2}{5}$ grams of zinc to every gram of copper.
Weight of zinc is $\frac{2}{5} \times 40\,\text{g} = 16\,\text{g}$.
Weight of bronze is $(40 + 16)\,\text{g} = 56\,\text{g}$.

The second type of problem is sharing a quantity in a given ratio. For example, to share £60 in the ratio 3:2, first add the 3 and 2 to obtain 5, the number of parts into which the £60 is divided. Each part is worth $£(60 \div 5) = £12$. The larger share is $3 \times £12 = £36$ and the smaller share is $2 \times £12 = £24$. The sum of the two shares is, of course, £60.

As a fraction of the total amount, the larger share is $\frac{36}{60} = \frac{3}{5}$, in its simplest form, and the smaller share is $\frac{24}{60} = \frac{2}{5}$. You can use fractions to divide a quantity in a given ratio.

**Example 19.3.2**
A company has 63 employees. The ratio of men to women is 4:5.
Find   (a) the number of men,   (b) the number of women.

(a) $\frac{4}{4+5}$, that is, $\frac{4}{9}$ of the employees are men.
So the number of men is $\frac{4}{9} \times 63 = \frac{4}{9} \times \frac{63}{1} = 28$. (See Section 3.8.)

(b) $\frac{5}{4+5}$, that is, $\frac{5}{9}$ of the employees are women.
So the number of women is $\frac{5}{9} \times 63 = \frac{5}{9} \times \frac{63}{1} = 35$.

You can find the number of women by subtracting the number of men from 63. The disadvantage of using this method is that, if your first answer is wrong, so is your second answer. It is better to evaluate each share and then check that the sum of the shares is correct.

You can use both methods to divide a quantity in a ratio which has more than two numbers. For example, to share 72 kg in the ratio 3:4:5 using the first method, add the 3, 4 and 5 to obtain 12, the number of parts. Each part is worth $72\,\text{kg} \div 12 = 6\,\text{kg}$, so the shares are $3 \times 6\,\text{kg}$, $4 \times 6\,\text{kg}$ and $5 \times 6\,\text{kg}$, that is, 18 kg, 24 kg and 30 kg. Alternatively, you can use the fractions $\frac{3}{3+4+5}$, $\frac{4}{3+4+5}$ and $\frac{5}{3+4+5}$, that is, $\frac{3}{12}$ or $\frac{1}{4}$, $\frac{4}{12}$ or $\frac{1}{3}$ and $\frac{5}{12}$.

## EXERCISE 19.3

1 Bronze is an alloy containing copper and tin in the ratio 3:1.
   (a) What weight of copper is mixed with 50 g of tin?
   (b) What weight of tin is mixed with 180 g of copper?
   (c) What weight of bronze contains 480 g of copper?
   (d) Work out the weight of copper and the weight of tin in 800 g of bronze.

2 An orange drink is made by mixing orange juice and water in the ratio 1:5.
   (a) How much water is mixed with 3 litres of orange juice?
   (b) How much water is in 24 litres of orange drink?

3 The ratio of Kevin's age to Michelle's age is 2:3. Kevin is 24 years old. How old is Michelle?

4 The ratio of the length of a rectangle to its width is 5:4. Its length is 55 cm. Work out its width.

5 The lengths of the sides of a triangle are in the ratio 2:3:4. The shortest side is 14 cm long. Find the lengths of the other two sides.

6 Anna and Brian share £80 in the ratio 7:3. Work out the amount each of them receives.

7 Cheryl and David share $420 in the ratio 2:5. Work out the amount each of them receives.

8 The ratio of Debbie's age to her father's age is 3:5. The sum of their ages is 96 years. Work out Debbie's age.

9 The angles of a triangle are in the ratio 6:5:7. Work out the size of each of its angles.

10 The perimeter of a triangle is 240 cm. The lengths of the sides are in the ratio 3:4:5. Work out the length of the longest side.

## 19.4 Direct proportion

If you walk at a steady speed of 4 mph, you will cover 4 miles in 1 hour, 8 miles in 2 hours, 12 miles in 3 hours and so on. If you walk for 5 hours, your distance–time graph will look like Figure 19.1.

The distance you walk obviously depends on the time, but there is a more precise relationship between the two

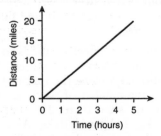

**Figure 19.1**

quantities. For example, you walk twice as far in 2 hours as you walk in 1 hour; you walk three times as far in 3 hours as you walk in 1 hour. When two quantities are related like this, they are said to be **directly proportional** (or proportional) to each other. If two quantities are directly proportional to each other, the graph showing the relationship between them is a straight line through the origin.

There are many examples of direct proportion in everyday life. The cost of petrol at a pump is directly proportional to the quantity of petrol you put in your tank. The cost of running an electric fire is directly proportional to the time for which it is on. If you are exchanging pounds for dollars, the number of dollars you get is directly proportional to the number of pounds you exchange.

There are two methods for answering questions on direct proportion. One is the **unitary** method, in which you find the value of one unit of the quantity. For example, if 8 cereal bars weigh 208 g, and you have to find the weight of 5 of these bars, the first step is to find the weight of one bar. To do this, you divide 208 g by 8, which gives 26 g. To find the weight of 5 bars, you then multiply 26 g by 5, which is 130 g.

The other method is the **multiplier** method. To answer the above example using this method, you find $\frac{5}{8}$ of 208 g, that is, $\frac{5}{8} \times \frac{208}{1} = 130$ g. Alternatively, you can convert $\frac{5}{8}$ to a decimal (0.625) and work out $0.625 \times 208$.

---

**Example 19.4.1**

$4\,\text{cm}^3$ of tin weigh 29.2 g. Using (a) the unitary method and (b) the multiplier method, work out the weight of $9\,\text{cm}^3$ of tin.

(a) Weight of $1\,\text{cm}^3$ of tin $= 29.2\,\text{g} \div 4 = 7.3\,\text{g}$.
Weight of $9\,\text{cm}^3$ of tin $= 9 \times 7.3\,\text{g} = 65.7\,\text{g}$.

(b) $9\,\text{cm}^3$ of tin weigh more than $4\,\text{cm}^3$ of tin.
So the multiplier is $\frac{9}{4}$ or 2.25.
Weight of $9\,\text{cm}^3$ of tin $= 2.25 \times 29.2\,\text{g} = 65.7\,\text{g}$.

---

## EXERCISE 19.4

1 A car travels 43 miles on one gallon of petrol. How far will it travel on 7 gallons?

2  £1 is worth 1.95 US dollars. How many dollars is £9 worth?

3  A motorist drives at a steady speed and goes 144 miles in 3 hours.
(a)  What distance will she cover in 5 hours?
(b)  How long will it take her to cover 120 miles?

4  5 litres of paint covers an area of 60 m².
(a)  What area will 15 litres of paint cover?
(b)  How much paint is needed to cover an area of 240 m²?

5  An electric fire uses 10 units of electricity in 4 hours.
(a)  How many units does it use in 3 hours?
(b)  For how long will it run on 17.5 units?

6  3 cm³ of aluminium weigh 8.1 g. Work out the weight of 10 cm³ of aluminium.

7  A recipe for 6 people requires 300 g of sugar and 1.5 kg of flour. Work out the weight of sugar and the weight of flour required for
(a)  12 people,   (b)  9 people,   (c)  4 people.

8  20 kg of cement is mixed with 60 kg of sand to make mortar.
(a)  What weight of sand should be mixed with 16 kg of cement?
(b)  What weight of cement should be mixed with 72 kg of sand?

9  There are 330 calories in 420 g of baked beans. How many calories are there in 700 g of baked beans?

10  5000 Japanese yen is worth £21.
(a)  How many pounds is 3000 yen worth?
(b)  To the nearest yen, how many yen is £40 worth?

## 19.5  Inverse proportion

The table shows the time taken to travel 120 miles at various steady speeds.

| Speed (mph) | 10 | 20 | 30 | 40 | 50 | 60 | 80 |
|---|---|---|---|---|---|---|---|
| Time (hours) | 12 | 6 | 4 | 3 | 2.4 | 2 | 1.5 |

When, for example, when a speed of 10 mph is multiplied by 4, becoming 40 mph, the time of 12 hours is divided by 4. Similarly, when a time of 6 hours is divided by 3, becoming 2 hours, the speed of 20 mph is multiplied by 3. When two quantities are related like this, they are said to be **inversely proportional** to each other.

The graph in Figure 19.2 shows the information from the table. Unlike direct proportion, when two quantities are inversely proportional to each other, an increase in one of the quantities causes a decrease in the other quantity.

As with direct proportion, you can answer questions on inverse proportion using either the unitary method or the multiplier method.

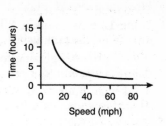

**Figure 19.2**

For example, if 3 men build a wall in 8 days, and you have to find how long 4 men would take, the first step using the unitary method is to work out how long one man would take, that is, $3 \times 8$ days $= 24$ days.

To find how long 4 men would take, you then divide 24 by 4 to obtain the answer of 6 days.

If you use the multiplier method for this question, the multiplier is $\frac{3}{4}$, not $\frac{4}{3}$, as 4 men will take less time than 3 men. Thus, 4 men take $\frac{3}{4}$ of 8 days, which is 6 days.

---

**Example 19.5.1**

A field provides grazing for 20 cows for 6 days. Using
(a) the unitary method and  (b) the multiplier method, find the number of days grazing the field would provide for 15 cows.

(a) One cow could graze for $6 \times 20$ days $= 120$ days.
So 15 cows would have grazing for $\frac{120}{15} = 8$ days.

(b) The field will provide 15 cows with grazing for more than 6 days.
So the multiplier is $\frac{20}{15}$ or $\frac{4}{3}$.
So 15 cows would have grazing for $\frac{4}{3} \times 6$ days $= 8$ days.

---

### EXERCISE 19.5

1 It takes 8 men 9 hours to dig a trench. How long would it take 6 men to dig the trench?

2 It takes 12 people 15 days to harvest a crop of raspberries. How long would it take 18 people?

3 Using 12 taps a tank is filled in 8 hours. How long would it take to fill the tank using 16 taps?

4 A field provides grazing for 18 sheep for 8 days. How many days grazing would it provide for 24 sheep?

5 It takes an hour to mow a lawn using a mower with blades 14 inches wide. How long would it take using a mower with blades 12 inches wide?

# 20

# Pythagoras' theorem and trigonometry

**In this chapter you will learn:**

- how to use Pythagoras' theorem to find lengths
- how to use Pythagoras' theorem to solve problems
- how to use trigonometry (tangents, sines and cosines) to find lengths and angles
- how to use trigonometry to solve problems.

## 20.1 Pythagoras' theorem

Pythagoras' theorem applies to **right-angled** triangles. The longest side of a right-angled triangle, which is opposite the right angle, is called the **hypotenuse**.

Pythagoras' theorem states that,

> **In a right-angled triangle, the area of the square drawn on the hypotenuse is equal to the sum of the areas of the squares drawn on the other two sides.**

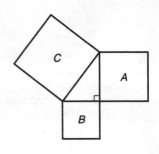

**Figure 20.1**

In Figure 20.1, this becomes area of square $C$ = area of square $A$ + area of square $B$.

This result was known to the Babylonians over 4000 years ago and, in Ancient Egypt, surveyors known as 'rope stretchers' knew that a triangle with sides of length 3, 4 and 5 units has a right angle between the sides of length 3 and 4. They used a knotted rope for marking right angles on the ground.

Pythagoras (c 580–500 BC), or one of his followers, may have been the first to prove the result. He settled in Crotona, a Greek colony in southern Italy, in about 530 BC. There he founded a mathematical and philosophical community which made major contributions in many areas of mathematics, especially number, to which they attached mystical qualities, and the harmonics of stringed instruments.

A proof of the theorem is in Section 20.3: the next section shows how to use the result.

## 20.2 Using Pythagoras' theorem

If you know the areas of the squares drawn on two sides of a right-angled triangle, you can use Pythagoras' theorem to find the area of the square drawn on the third side.

For example, in Figure 20.2, if the area of square $A$ is $20\,\text{cm}^2$ and the area of square $B$ is $13\,\text{cm}^2$, then the area of square $C$ is $(20 + 13)\,\text{cm}^2 = 33\,\text{cm}^2$.

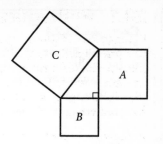

**Figure 20.2**

In practice, when you use Pythagoras' theorem, it will usually be to find the length of the third side of a right-angled triangle, if you know the lengths of the other two sides. For this purpose, it is more useful if the theorem is stated in terms of the lengths of the sides. (See Figure 20.3.)

$$c^2 = a^2 + b^2.$$

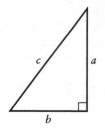

**Figure 20.3**

In Figure 20.4, the value of $a$ is 8 and the value of $b$ is 6, the units being centimetres. As $x$ takes the place of $c$, using Pythagoras' theorem gives

$$x^2 = 8^2 + 6^2 = 64 + 36 = 100.$$

As $x$ is positive and $x^2 = 100$, $x = 10$, so the third side is $10\,\text{cm}$.

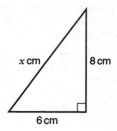

**Figure 20.4**

As the square root of a square number is a whole number, it was easy to find $x$, but in other examples you must use your calculator to find square roots. You should give your result to an appropriate degree of accuracy, but three significant figures will usually suffice. Example 20.2.1 illustrates this point.

### Example 20.2.1

Calculate the length marked $x$ cm in the right-angled triangle, giving your answer correct to three significant figures.

From Pythagoras' theorem,

$$7^2 = x^2 + 4^2$$
$$49 = x^2 + 16$$
$$x^2 = 49 - 16 = 33$$
$$x = \sqrt{33} = 5.74, \text{ correct to three significant figures.}$$

The length is $5.74$ cm, correct to three significant figures.

You can split isosceles and equilateral triangles into two right-angled triangles by a line of symmetry and then use Pythagoras' theorem.

### Example 20.2.2

The diagram shows an isosceles triangle with a base of 24 cm and a vertical height of 9 cm. Calculate the length of the sides marked $l$ cm. From Pythagoras' theorem in the right-angled triangle,

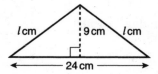

$$l^2 = 12^2 + 9^2 = 144 + 81 = 225 \quad \text{so} \quad l = \sqrt{225} = 15.$$
The length of the side is $15$ cm.

Although it will not be proved in this book, Pythagoras' theorem works in reverse. If the sum of the areas of squares on two of the sides of a triangle is equal to the area of the third square, then the triangle is right-angled. You will need this property in Exercise 20.1, Question 2.

## EXERCISE 20.1

1  Copy and complete the table, which refers to Figure 20.1.

| Area of square $A$ | Area of square $B$ | Area of square $C$ |
|:---:|:---:|:---:|
| 12 cm² | 7 cm² | |
| 9 cm² | | 26 cm² |
| | 13 cm² | 29 cm² |

2  Three squares have areas of 12 cm², 18 cm² and 30 cm². If these squares are arranged as in Figure 20.1, will the triangle in the centre be right-angled? Give a reason for your answer.

In the rest of this exercise, give your answers to three significant figures where necessary.

3  Calculate the length of each of the sides marked with a letter. The units are centimetres.

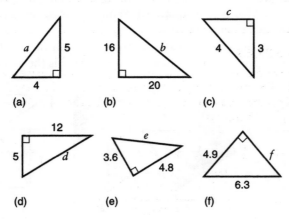

(a)    (b)    (c)

(d)    (e)    (f)

4  The sides of a square are 5 cm long. Calculate the length of a diagonal.

5  A rectangle is 10 cm long and 7 cm wide. Calculate the length of a diagonal.

6  In the isosceles triangle $ABC$, $AC = BC = 8$ cm. The vertical height, $CD$, of the triangle is 5 cm. Calculate the length of the base, $AB$.

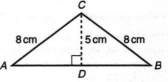

7 The sides of an equilateral triangle are 6 cm long. Calculate the vertical height of the triangle.

8 The diagram shows a trapezium *ABCD* with right angles at *B* and C. *AB* = 9 cm, *BC* = 4 cm and *CD* = 6 cm. Calculate the perimeter of the trapezium.

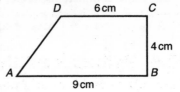

## 20.3 Proof of Pythagoras' theorem

This proof of Pythagoras' theorem is based on a dissection. Figure 20.5 shows a right-angled triangle with adjacent sides labelled $a, b$ and hypotenuse $c$.

**Figure 20.5**

The first diagram in Figure 20.6 shows four copies of the triangle arranged in the form of a square surrounding a smaller square. The area of the larger square is $c^2$.

If you look at the square in the centre, its side is of length $x$; from the dimensions in the figure, $x + a = b$. This is the same as $b - x = a$.

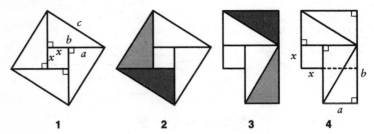

**Figure 20.6**

The second figure is identical to the first, but with two of the triangles shaded. The third figure shows how the two shaded triangles have been moved. The fourth figure is the same as the third, but with an additional dotted line drawn across it.

Note that the fourth figure is made up of the same shapes as the first, so its area is the same as the area of the first square, which is $c^2$.

Below the dotted line is a rectangular shape whose breadth is $a$, and whose height is $b - x$, which is the same as $a$, so its area is $a^2$.

Above the dotted line is a rectangular shape whose breadth is $b$, and whose height is $x + a$, which is the same as $b$, so its area is $b^2$.

So the area of the fourth figure is $a^2 + b^2$. So $a^2 + b^2 = c^2$.

## 20.4 Pythagoras' theorem problems

When solving problems on Pythagoras' theorem, always draw a diagram to show the information you are given. You may introduce a letter to represent the required distance but you should not use it in the final statement of the answer. Include units with your answer, if necessary.

---

**Example 20.4.1**

A ladder 6 m long leans against a wall so that the top of the ladder is 5 m from the ground. How far is the foot of the ladder from the wall? Give your answer correct to three significant figures.

Let $x$ m be the distance of the foot of the ladder from the wall.

Then, using Pythagoras' theorem,
$$6^2 = x^2 + 5^2.$$

So $x^2 = 36 - 25 = 11$,
giving $x = \sqrt{11} = 3.32$, correct to three significant figures.

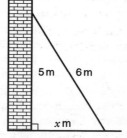

The foot of the ladder is 3.32 m, correct to three significant figures, from the wall.

---

### EXERCISE 20.2

Give your answers to three significant figures where necessary.

1  The foot of a ladder is 2 m from a wall and the top of the ladder is 6 m above the ground. Calculate the length of the ladder.

2  A ladder 7 m long leans against a wall so that the foot of the ladder is 2 m from the wall. How far is the top of the ladder above the ground?

3  Oxford is 48 miles South of Birmingham and 19 miles East of Birmingham. Calculate the straight line distance between Oxford and Birmingham.

4 The lengths of the diagonals of a rhombus are 16 cm and 12 cm. Calculate the length of the sides of the rhombus. (Hint: the diagonals of a rhombus bisect each other at right angles.)

5 The sides of a rhombus are 8 cm long. One of its diagonals is 8 cm long. Calculate the length of the other diagonal.

6 The lengths of two of the sides of a right-angled triangle are 4 cm and 5 cm. Calculate the **two** possible lengths of the third side.

7 The diagonals of a square are 6 cm long. Calculate the length of the sides of the square.

8 *A* and *B* are points with coordinates (2,4) and (6,7) respectively. The units are centimetres. Calculate the distance *AB*.

## 20.5 Trigonometry

Trigonometry is the branch of mathematics about the relationships between the sides and angles of triangles. The word 'trigonometry' is derived from the Greek *trigonon* (triangle) and *metron* (measure).

Trigonometry was used in navigation, surveying and astronomy, often when the problem was to find an inaccessible distance, such as that between the Earth and the moon. The origins of trigonometry are probably the works of the Greek astronomer and mathematician Hipparchus (c 190–120 BC).

Thereafter, the development of trigonometry had many strands, initially Indian and Arab and, later, European. The first major Western European work on trigonometry *De trinaglis omni modis* was written by the German mathematician Regiomontanus in 1464 and, in the seventeenth century, through the improvements in algebraic symbolism, trigonometry became the analytic science it is today.

## 20.6 The tangent ratio

Figure 20.7 shows a right-angled triangle *ABC*, in which $\angle A = 35°$. The length, *AB*, of the side **adjacent** to the 35° angle is drawn to be 2 cm. If you measure the length of the side *BC* **opposite** the 35° angle you will find that it is very close to 1.4 cm.

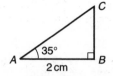

**Figure 20.7**

(Figure 20.7 is drawn accurately, so you can actually measure the length of BC on the diagram.)

If you enlarge the triangle in Figure 20.7 to get the triangle in Figure 20.8 so that the length of the side adjacent to the 35° angle is 3 cm, and you again measure the length of the side opposite to the 35° angle, you will find that the length is very close to 2.1 cm.

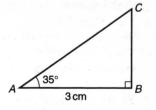

**Figure 20.8**

If you calculate the ratio $\dfrac{\text{opposite side}}{\text{adjacent side}} = \dfrac{BC}{AB}$ in each case, you will find that, in both cases, $\dfrac{BC}{AB} = 0.7$.

In fact, for all right-angled triangles with an angle of 35°, no matter what the length of the adjacent side, the ratio $\dfrac{\text{opposite side}}{\text{adjacent side}}$ is 0.7, correct to one decimal place.

The ratio $\dfrac{\text{opposite side}}{\text{adjacent side}}$, usually abbreviated to $\dfrac{\text{opposite}}{\text{adjacent}}$, is called the **tangent** of the angle 35°, written as **tan**. In the case above, tan 35° = 0.7.

In general, using the Greek letter $\theta$ (theta) to represent any angle, $\tan \theta = \dfrac{\text{opposite}}{\text{adjacent}}$. (See Figure 20.9.)

Remember that the sides 'opposite' and 'adjacent' must relate to the angle you are considering.

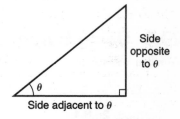

**Figure 20.9**

You must also appreciate that it could be argued that there are always two sides which are adjacent to a given angle. The **adjacent** is the side which is *not* the hypotenuse.

## 20.7 Values of the tangent

You can find the value of the tangent of an angle using your calculator. After making sure that it is in degree mode, use your calculator to check that tan 35° = 0.700 207.... You may need to consult your handbook.

You should also find that tan 45° = 1. Tangents continue to increase as the angle increases from 45° towards 90°, and tangents of angles close to 90° are very large. The ratio tan 90°, however, does not exist. If θ were 90° in Figure 20.9, you would no longer have a triangle, so tan 90° is not defined. Try finding tan 90° on your calculator; what happens?

If you know the tangent of an angle and want to find the angle, you use the **inverse tangent**. This is often written as $\tan^{-1}$, or sometimes as arctan. For example, if you need to know which angle has a tangent of 1.9, you should find that $\tan^{-1} 1.9 = 62.241\ldots°$. If you have problems, consult your calculator handbook.

### EXERCISE 20.3

1 Use your calculator to find the values, correct to three significant figures, of the tangents of the following angles.
   (a) tan 40°      (b) tan 65°      (c) tan 2°
   (d) tan 32.7°    (e) tan 0.7°     (f) tan 74.83°
   (g) tan 89.9°    (h) tan 89.99°

2 Use your calculator to find the values of θ, giving your answers correct to one decimal place.
   (a) $\tan\theta = 0.6$      (b) $\tan\theta = 3$       (c) $\tan\theta = 0.25$
   (d) $\tan\theta = 5$        (e) $\tan\theta = 0.08$    (f) $\tan\theta = 0.893$
   (g) $\tan\theta = 9.372$    (h) $\tan\theta = 23.4$

## 20.8  Using tangents

You can use tangents to calculate angles and lengths in right-angled triangles.

**Example 20.8.1**

Calculate the size of the angle marked θ in the diagram. Give your answer correct to one decimal place.

$$\tan\theta = \frac{\text{opposite}}{\text{adjacent}} = \frac{4.3}{3.7}$$

$$= 1.1621\ldots$$

$$\theta = \tan^{-1} 1.1621\ldots = 49.3°,$$
correct to one decimal place.

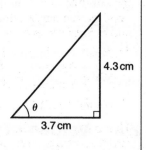
4.3 cm
3.7 cm

The triangle may sometimes be in a position where identifying the opposite and adjacent sides to the relevant angle takes some care.

---

**Example 20.8.2**

Calculate the length of the side marked $a$ cm in the Figure. Give your answer correct to three significant figures.

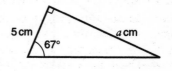

If you focus on the angle 67°, you will see that $a$ cm is opposite the 67° angle and the 5 cm side is adjacent to it.

$$\tan 67° = \frac{\text{opposite}}{\text{adjacent}} = \frac{a}{5}, \text{ so } \tan 67° = \frac{a}{5}.$$

Looking at this as an equation and multiplying both sides by 5 gives

$$a = 5\tan 67° = 5 \times 2.3558\ldots = 11.779\ldots.$$

The length of the side is 11.8 cm, correct to three significant figures.

---

You can also use tangents to solve problems.

---

**Example 20.8.3**

A flag pole is 6 m high. A guy rope is attached to the top of the flag pole and to a peg in the ground 3.5 m from the foot of the flag pole. Calculate the angle the guy rope makes with the ground. Give your answer correct to one decimal place.

Let $\theta$ be the angle which the guy rope makes with the ground.
Then

$$\tan \theta = \frac{\text{opposite}}{\text{adjacent}} = \frac{6}{3.5} = 1.7142\ldots.$$

$$\theta = \tan^{-1} 1.7142\ldots = 59.743\ldots°.$$

The angle which the guy rope makes with the ground is 59.7°, (correct to one decimal place).

---

# EXERCISE 20.4

1  Find the marked angles or sides, giving lengths correct to three
significant figures and angles correct to one decimal place.

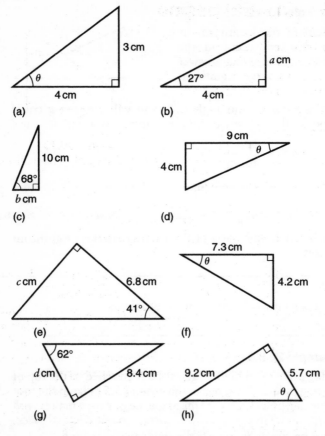

2  A ladder resting against a wall makes an angle of 63° with the
ground. The foot of the ladder is 4.7 m from the wall. Calculate
the height of the top of the ladder above the ground.

3  The lengths of two adjacent sides of a rectangle are 9 cm and
7 cm. Calculate the size of the angle a diagonal of the rectangle
makes with a 9 cm side.

4  The bearing of York from Leeds is 058°. York is 18 miles East
of Leeds. Calculate the distance York is North of Leeds.

5  A man 1.8 m tall stands 100 m away from the foot of the Eiffel
Tower on level ground. To see the top, he has to look up at an

angle of 71.5° above the horizontal. (This is called an **angle of elevation**.) Calculate the height of the Eiffel Tower.

## 20.9  Sine and cosine

Figure 20.10 shows a right-angled triangle. The side $a$ is opposite angle $\theta$; $b$ is adjacent to $\theta$ and $c$ is the hypotenuse. You already know that, for a fixed angle $\theta$, the ratio $\dfrac{a}{b}$ is constant (and equal to $\tan \theta$).

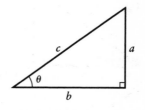

**Figure 20.10**

The ratio $\dfrac{a}{c}$, that is, $\dfrac{\text{opposite}}{\text{hypotenuse}}$, is also constant. This ratio is the **sine** of the angle $\theta$ and is written $\sin \theta$.

Similarly, the ratio $\dfrac{b}{c}$, that is, $\dfrac{\text{adjacent}}{\text{hypotenuse}}$, is constant. This ratio is the **cosine** of the angle $\theta$ and is written $\cos \theta$.

Thus:

$$\sin \theta = \frac{\text{opposite}}{\text{hypotenuse}} = \frac{a}{c}; \qquad \cos \theta = \frac{\text{adjacent}}{\text{hypotenuse}} = \frac{b}{c}.$$

Finding the value of the sine or cosine of an angle using your calculator is similar to finding the tangent of an angle. Sines and cosines, however, cannot be greater than 1. You can use $\sin^{-1}$ and $\cos^{-1}$ to find the inverse sine and the inverse cosine in the same way that you used $\tan^{-1}$ to find the inverse tangent.

You can use sines and cosines to find angles and lengths in right-angled triangles in a way similar to the one used for tangents.

For example, in Figure 20.11, the 7 cm side is opposite angle $\theta$ and the 8 cm side is the hypotenuse. The ratio relating the opposite side and the hypotenuse is sine. So,

$$\sin \theta = \frac{\text{opposite}}{\text{hypotenuse}} = \frac{7}{8} = 0.875.$$

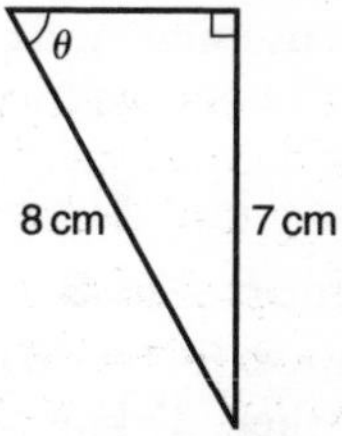

Then using your calculator to find $\sin^{-1} 0.875$ gives $\theta = 61.04449\ldots^\circ$.

**Figure 20.11**

Thus, correct to one decimal place, $\theta = 61.0^\circ$.

### Example 20.9.1

Calculate the length of the side marked $a$ m in the figure. Give your answer correct to three significant figures.

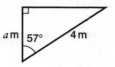

$$\cos 57° = \frac{\text{adjacent}}{\text{hypotenuse}} = \frac{a}{4},$$

so $a = 4\cos 57° = 4 \times 0.5446... = 2.1785....$

Thus the length is 2.18 m, correct to three significant figures.

### Example 20.9.2

Calculate the length of the side marked $b$ cm in the figure. Give your answer correct to three significant figures.

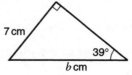

$$\sin 37° = \frac{\text{opposite}}{\text{hypotenuse}} = \frac{7}{b}, \text{ so } \sin 37° = \frac{7}{b}.$$

Multiplying both sides by $b$ gives $b\sin 37° = 7$, so, dividing by $\sin 37°$, gives $b = \dfrac{7}{\sin 37°} = \dfrac{7}{0.6018...} = 11.6314....$

Thus the length is 11.6 cm, correct to three significant figures.

You can use sines and cosines to solve problems. As with tangents, always draw a diagram.

### Example 20.9.3

An aircraft flies 50 km in a straight line. It is then 37 km North of its starting point. Find the bearing on which it flew. Give your answer correct to the nearest degree. In the figure, let $\theta$ be the required bearing.

$$\cos \theta = \frac{\text{adjacent}}{\text{hypotenuse}} = \frac{37}{50} = 0.74.$$

$$\theta = \cos^{-1} 0.74 = 42.26\ldots°.$$

The aircraft flew on the bearing 042°, correct to the nearest degree.

## EXERCISE 20.5

1 Find the marked angles or sides. Give lengths correct to three significant figures and angles correct to one decimal place.

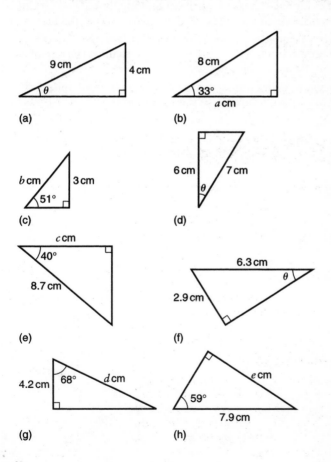

2 An aircraft flies for 400 m, climbing at an angle of 9° above the horizontal. Calculate its increase in height.

3 A boy flies a kite with a piece of string 25 m long. Calculate the size of the angle the string makes with the horizontal when the kite is 20 m above the ground.

4 A ladder leans against a wall. It makes an angle of 63° with the ground and the foot of the ladder is 3 m from the wall. Calculate the length of the ladder.

5 A wire rope is attached to the top of a TV mast and to a point on the ground. The length of the rope is 183 m and the height of the mast is 128 m. Calculate the angle between the rope and the ground.

6 The diagonal of a rectangle is 10 cm long. It makes an angle of 36° with one of the sides. Calculate the area of the rectangle.

7 The two equal sides of an isosceles triangle are 7.5 cm long. The length of the other side is 10 cm. Calculate the size of each of the angles of the triangle.

8 A regular pentagon is drawn inside a circle of radius 8 cm. Calculate the perimeter of the pentagon.

**21**

## indices and standard form

**In this chapter you will learn:**

- how to use indices and the laws of indices
- how to express a number as the product of powers of its prime factors
- how to find the highest common factor (HCF) of two numbers
- how to find the lowest common multiple (LCM) of two numbers
- how to use standard form.

## 21.1 Indices

You saw in Chapter 11 how indices are a shorthand for repeated multiplication. For example, you can write $a \times a \times a \times a \times a$ as $a^5$, read as '$a$ to the power 5', where 5 is the index or power.

In a similar way, you can use indices (the plural of index) to express powers of numbers. For example, $2^5$ is shorthand for $2 \times 2 \times 2 \times 2 \times 2$, the value of which, when you carry out the multiplications, is 32.

You may be able to evaluate $2^5$ using your calculator without carrying out repeated multiplication: some calculators use a $y^x$ key; others use a key marked $\wedge$.

### EXERCISE 21.1

1 Write the following in index form.
   (a) $5 \times 5$     (b) $7 \times 7 \times 7$     (c) $2 \times 2 \times 2 \times 2$

2 Find the value of each of the following.
   (a) $3^2$     (b) $5^3$     (c) $2^4$     (d) $10^6$

3 Find the values of $x$ which satisfy the following equations.
   (a) $2^x = 8$     (b) $3^x = 81$     (c) $10^x = 10\,000$

4 Find the values of the following.
   (a) $2^3 \times 3^2$     (b) $2^4 \times 5^2$     (c) $2^2 \times 3^4 \times 5$

## 21.2 Laws of indices

In Chapter 11, you used the laws of indices to simplify algebraic expressions. For convenience, some of the laws are restated below.

$$a^m \times a^n = a^{m+n} \qquad a^m \div a^n = a^{m-n}$$
$$(a^m)^n = a^{mn} \qquad a^1 = a$$

You can, for example, use the first law to express $2^3 \times 2^5$ as a single power of 2. You add the indices to obtain $2^3 \times 2^5 = 2^{3+5} = 2^8$.

---

**Example 21.2.1**

Express each of the following as a single power of 3.

(a) $3^7 \div 3^2$     (b) $3^8 \div 3$     (c) $(3^5)^2$

(a) Subtracting the indices gives $3^7 \div 3^2 = 3^{7-2} = 3^5$.
(b) Write 3 as $3^1$. Then $3^8 \div 3 = 3^8 \div 3^1 = 3^{8-1} = 3^7$.
(c) Multiplying the indices gives $(3^5)^2 = 3^{5 \times 2} = 3^{10}$.

You cannot use the laws of indices to express expressions like $2^5 \times 3^4$ as a single power of a number, because 2 and 3 are different numbers.

### EXERCISE 21.2

**1** Write each of the following as a single power of the number.

(a) $2^4 \times 2^5$      (b) $5^8 \div 5^2$      (c) $(3^4)^2$

(d) $10^3 \times 10$      (e) $5^9 \div 5$      (f) $2^5 \times 2^3 \times 2^4$

(g) $3^7 \times 3^4 \div 3^5$      (h) $\dfrac{10^4 \times 10^8}{10^7}$

## 21.3 Prime factors

The factors of 20 are 1, 2, 4, 5, 10 and 20. Of these factors, 2 and 5 are **prime** numbers, so 2 and 5 are called the **prime factors** of 20. Every number can be expressed as the product of powers of its prime factors. You can express 20, for example, as $2^2 \times 5$. In a simple case like this, you might be able to do the working in your head but in more complicated cases it is easier to use repeated division.

The first prime number is 2, which is a factor of 20. You can divide 20 exactly by 2 twice.

$20 \div 2 = 10$ and $10 \div 2 = 5$.

The second prime number is 3 but 3 is not a factor of 5.

The third prime number is 5, which is a factor of 5.

$5 \div 5 = 1$, which completes the dividing.

You can set out the working like this.

| 2 | 20 |
|---|----|
| 2 | 10 |
| 5 | 5  |
|   | 1  |

So $20 = 2 \times 2 \times 5 = 2^2 \times 5$.

The next example illustrates this method with a larger number.

### Example 21.3.1

Express 360 as the product of powers of its prime factors.

| 2 | 360 |
|---|-----|
| 2 | 180 |
| 2 | 90  |
| 3 | 45  |
| 3 | 15  |
| 5 | 5   |
|   | 1   |

So $360 = 2 \times 2 \times 2 \times 3 \times 3 \times 5 = 2^3 \times 3^2 \times 5$.

### EXERCISE 21.3

1  Write these numbers as products of powers of prime factors.

(a) 12      (b) 18      (c) 36      (d) 45      (e) 48
(f) 60      (g) 108     (h) 225     (i) 280     (j) 2200

## 21.4  Highest common factor

The **highest common factor** (HCF) of two numbers is the largest number which will divide exactly into both of them.

The factors of 24 are 1, 2, 3, 4, 6, 8, 12, 24.

The factors of 36 are 1, 2, 3, 4, 6, 9, 12, 18, 36.

Six numbers divide exactly into both 24 and 36: 1, 2, 3, 4, 6 and 12.

The highest factor 24 and 36 have in common is 12, so 12 is the highest common factor of 24 and 36.

In this case, the lists of factors are short and you may be able to find the HCF mentally. However, to find the HCF of larger numbers such as 540 and 1008, listing the factors would be tedious, so it would be useful to have a way of finding the HCF more directly.

In prime factor form, $24 = 2^3 \times 3, 36 = 2^2 \times 3^2$. If you look at the powers of 2, you see that $2^2$ is the highest power of 2 which divides both 24 and 36. Similarly, 3 is the highest power of 3 which divides both of them. So $2^2 \times 3 = 12$ is the largest number which divides both 24 and 36. In general, the HCF of two numbers is the product of the **lowest** powers of the prime factors they have in common. You can use this method to find the HCF of more than two numbers.

---

**Example 21.4.1**

Find in index form the HCF of $2^3 \times 3^2 \times 5^4$ and $2^4 \times 5^2 \times 7^3$.

The common prime factors are 2 and 5.

The lowest power of 2 which divides both is $2^3$ and the lowest power of 5 which divides both is $5^2$, so the HCF is $2^3 \times 5^2$.

---

**Example 21.4.2**

Find the HCF of 540 and 1008.

In prime factor form, $540 = 2^2 \times 3^3 \times 5$ and $1008 = 2^4 \times 3^2 \times 7$.

The common prime factors are 2 and 3.

The lowest power of 2 which divides both is $2^2$ and the lowest power of 3 which divides both is $3^2$. So the HCF is $2^2 \times 3^2 = 36$.

---

### EXERCISE 21.4

1 Write down the highest common factor of the following pairs of numbers.
   (a) 6 and 9  (b) 14 and 21  (c) 18 and 24  (d) 20 and 30

2 Write down in index form the highest common factor of the following sets of numbers.
   (a) $2^3 \times 3^2, 2^2 \times 3^4$  (b) $2^4 \times 3^3 \times 5^2, 2^3 \times 3^5 \times 7^2$
   (c) $2 \times 3^2 \times 5^3, 2^3 \times 5^2 \times 11$ (d) $2^4 \times 5, 2^3 \times 5^4, 2^5 \times 5^2 \times 7$

3 Find the highest common factor of the following sets of numbers.
   (a) 120 and 168   (b) 300 and 420   (c) 120, 280 and 360

## 21.5 Lowest common multiple

The **lowest common multiple** (LCM) of two numbers is the lowest number which is a multiple of the two numbers.

The multiples of 12 are 12, 24, 36, 48, 60, 72, 84, 96, 108, 120, ....

The multiples of 15 are 15, 30, 45, 60, 75, 90, 105, 120, 135, ....

The multiples which 12 and 15 have in common are 60, 120, ....

The lowest multiple 12 and 15 have in common is 60, so 60 is the lowest common multiple of 12 and 15.

In prime factor form, $12 = 2^2 \times 3, 15 = 3 \times 5$. The highest powers of 2, 3 and 5 respectively are $2^2$, 3 and 5, so both 12 and 15 divide into $2^2 \times 3 \times 5 = 60$. Thus, 60 is the LCM of the two numbers. In general, the LCM of two numbers is the product of the highest powers of all their prime factors, not just the ones they have in common.

---

**Example 21.5.1**

Write down in index form the lowest common multiple of $2^3 \times 5^2 \times 7$ and $2^2 \times 3^4 \times 5^3$.

The highest power of each of the prime factors is $2^3$, $3^4$, $5^3$ and 7. The LCM is $2^3 \times 3^4 \times 5^3 \times 7$.

---

**Example 21.5.2**

Find the LCM of 45 and 60.

In prime factor form, $45 = 3^2 \times 5$ and $60 = 2^2 \times 3 \times 5$.

The highest power of each of the prime factors is $2^2$, $3^2$ and 5.

In index form, the LCM is $2^2 \times 3^2 \times 5 = 180$.

---

## EXERCISE 21.5

1 Write down the LCM of the following sets of numbers.
(a) 3 and 5 (b) 4 and 6 (c) 8 and 12 (d) 2, 3 and 4

2 Write down in index form the LCM of the following numbers.
(a) $2^3 \times 3^4, 2^2 \times 3^5$ (b) $2^4 \times 3^2 \times 5, 2^5 \times 3 \times 7^3$
(c) $2^3 \times 5^2, 2^6 \times 5 \times 11^2$ (d) $2^4 \times 3^3, 3^2 \times 5^3, 2^3 \times 5 \times 7$

3 Find the lowest common multiple of the following numbers.
(a) 40 and 50 (b) 24 and 30 (c) 18, 20 and 27

# 21.6 Standard form – large numbers

In this section, you will be using powers of 10. Here are some of them.

$$10^4 = 10 \times 10 \times 10 \times 10 = 10\,000$$
$$10^3 = 10 \times 10 \times 10 \qquad = 1000$$
$$10^2 = 10 \times 10 \qquad\qquad = 100$$
$$10^1 = 10 \qquad\qquad\qquad = 10$$

Note that the value of $10^4$ is 1 followed by 4 zeros; the value of $10^3$ is 1 followed by 3 zeros and so on. You can extend this to higher powers of 10, so $10^7$ is 1 followed by 7 zeros, that is, $10^7 = 10\,000\,000$.

The value of $2 \times 10^3$ is $2 \times 1000 = 2000$. Again, the index is the same as the number of zeros.

The number $2 \times 10^3$ is an example of a number written in **standard form,** or **standard index form.**

Standard form consists of a number between I and 10 multiplied by a power of 10. The number between 1 and 10 need not be a whole number; the number $7.1 \times 10^4$ is in standard form. The value of $7.1 \times 10^4$ is $7.1 \times 10\,000$, which is $71\,000$.

To write a number in standard form, write down the number between 1 and 10 and then decide what power of 10 it has to be multiplied by. With $657\,000$ the number between 1 and 10 is 6.57, which has to be multiplied by $100\,000$, which is $10^5$. So $657\,000 = 6.57 \times 10^5$.

Standard form is often used to write very large numbers. For example, the distance from Earth to the star Sirius is $50\,000\,000\,000\,000$ miles. It is much more convenient to express this distance in standard form as $5 \times 10^{13}$ miles, rather than count all those zeros.

---

**Example 21.6.1**

Write in standard form
(a) 5730   (b) 600 000   (c) 2.8 million.
(a) $5730 = 5.73 \times 1000$, so $5730 = 5.73 \times 10^3$.
(b) $600\,000 = 6 \times 100\,000 = 6 \times 10^5$.
(c) 2.8 million $= 2\,800\,000 = 2.8 \times 1\,000\,000 = 2.8 \times 10^6$.

### Example 21.6.2
Write $6.47 \times 10^4$ as an ordinary number.
$6.47 \times 10^4 = 6.47 \times 10\,000 = 64\,700$.

Try using your calculator to check the answers to the last two examples. You may have to use the manual to find out how to do it. The number of figures on your calculator display will limit the size of the number it can show as an ordinary number.

### EXERCISE 21.6

1 Write the following numbers in standard form.
(a) 900      (b) 3000      (c) 700 000 (d) 40 000 000
(e) 32 000    (f) 170 000    (g) 4520     (h) 275 000 000
(i) 52 million (j) 4.6 million

2 Write the following numbers as ordinary numbers.
(a) $5 \times 10^2$   (b) $8 \times 10^4$   (c) $9 \times 10^7$   (d) $8.7 \times 10^5$
(e) $9.1 \times 10^3$ (f) $2.3 \times 10^6$ (g) $1.97 \times 10^4$ (h) $4.21 \times 10^{10}$

3 Light travels 300 000 000 metres in one second. Write this number in standard form.

4 The age of the Sun is approximately 7500 million years. Write this number in standard form.

5 The average distance of the Earth from the Sun is given as $9.37 \times 10^7$ miles. Write this as an ordinary number.

## 21.7 Standard form – small numbers

Here are some numbers expressed in standard form.

$$37\,000 = 3.7 \times 10^4$$
$$3700 = 3.7 \times 10^3$$
$$370 = 3.7 \times 10^2$$
$$37 = 3.7 \times 10^1$$

You can obtain each number on the left by dividing the one above it by 10 and each index is one less than the one above it. You can use these relationships to continue the pattern.

$$3.7 = 3.7 \times 10^0$$
$$0.37 = 3.7 \times 10^{-1}$$
$$0.037 = 3.7 \times 10^{-2}$$
$$0.0037 = 3.7 \times 10^{-3}$$

This shows that, when you express a number less than 1 in standard form, the power of 10 is negative. One way of finding the index is to add 1 to the number of zeros on the right of the decimal point before the first significant figure and then make this number negative. For example, 0.00072 has 3 zeros on the right of the decimal point before the first significant figure, 7. Adding 1 gives 4 and so the index is $-4$. Thus, in standard form, 0.00072 is $7.2 \times 10^{-4}$.

---

**Example 21.7.1**

Write in standard form (a) 0.000 003   (b) 0.0912

(a) There are 5 zeros on the right of the decimal point.

To find the index, add 1 to 5 (6) and make it negative ($-6$).

Then $0.000\,003 = 3 \times 10^{-6}$.

(b) There is 1 zero on the right of the decimal point.

To find the index, add 1 to 1 (2) and make it negative ($-2$).

Then $0.0912 = 9.12 \times 10^{-2}$.

---

**Example 21.7.2**

Write $1.9 \times 10^{-5}$ as an ordinary number.

The index, $-5$, tells you that there are 4 zeros on the right of the decimal point before the 1.

Thus $1.9 \times 10^{-5} = 0.000\,019$.

---

Try using your calculator to check the answers to the last two examples.

You may need to consult your manual.

### EXERCISE 21.7

1 Write the following numbers in standard form.
   (a) 0.07    (b) 0.0004    (c) 0.3    (d) 0.000 000 09
   (e) 0.0084  (f) 0.000 029 (g) 0.007 91 (h) 0.739

2 Write the following numbers in standard form as ordinary numbers.
   (a) $4 \times 10^{-3}$  (b) $8 \times 10^{-5}$  (c) $9 \times 10^{-1}$  (d) $3 \times 10^{-9}$
   (e) $1.3 \times 10^{-2}$ (f) $9.7 \times 10^{-4}$ (g) $4.73 \times 10^{-6}$ (h) $5.08 \times 10^{-7}$

3 The wavelength of visible light is 0.0005 mm. Write this number in standard form.

4 The coefficient of linear expansion of copper is 0.000 016 9 per degree C. Write this number in standard form.

5 The lifetime in seconds of Mu Minus, an atomic particle, is $2.2 \times 10^{-6}$. Write this as an ordinary number.

# 21.8 Standard form calculations

You can do simple standard form calculations mentally with the laws of indices. For more awkward calculations, use your calculator.

---

**Example 21.8.1**

Work out $(3 \times 10^6) \times (5 \times 10^4)$. Give your answer in standard form.

You can change the order of the multiplication without affecting it.

So $(3 \times 10^6) \times (5 \times 10^4) = 3 \times 5 \times 10^6 \times 10^4 = 15 \times 10^{10}$.

But this is not in standard form; write 15 in standard form as $1.5 \times 10^1$.

Then $15 \times 10^{10} = 1.5 \times 10^1 \times 10^{10} = 1.5 \times 10^{11}$, so $(3 \times 10^6) \times (5 \times 10^4) = 1.5 \times 10^{11}$.

---

**Example 21.8.2**

Work out $\dfrac{8 \times 10^7}{2 \times 10^4}$.

You can write $\dfrac{8 \times 10^7}{2 \times 10^4}$ as $\dfrac{8}{2} \times \dfrac{10^7}{10^4}$.

Then $\dfrac{8 \times 10^7}{2 \times 10^4} = \dfrac{8}{2} \times \dfrac{10^7}{10^4} = 4 \times 10^3$.

---

**Example 21.8.3**

Work out $(5.4 \times 10^4) + (3.6 \times 10^3)$. Give your answer in standard form. In this case, you have to evaluate each of the terms first.

$5.4 \times 10^4 = 54\,000$ and $3.6 \times 10^3 = 3600$.

Adding the numbers gives $54\,000 + 3600 = 57\,600$.

In standard form, $57\,600 = 5.76 \times 10^4$.

So $(5.4 \times 10^4) + (3.6 \times 10^3) = 5.76 \times 10^4$

## EXERCISE 21.8

1 Work out the following without using a calculator. Give your answers in standard form.

(a) $(3 \times 10^5) \times (2 \times 10^6)$    (b) $(6 \times 10^3) \times (4 \times 10^4)$

(c) $\dfrac{9 \times 10^8}{3 \times 10^5}$    (d) $\dfrac{4 \times 10^6}{5 \times 10^2}$

(e) $(4 \times 10^4) + (3 \times 10^3)$    (f) $(5 \times 10^6) - (7 \times 10^4)$

(g) $(3.1 \times 10^5) + (4.7 \times 10^3)$   (h) $(6.3 \times 10^3) - (3.8 \times 10^2)$

2 Using a calculator, work out the following. Give your answers in standard form, correct to three significant figures.

(a) $(6.71 \times 10^3) \times (4.29 \times 10^6)$

(b) $\dfrac{9.7 \times 10^5}{3.8 \times 10^3}$

(c) $(4.82 \times 10^4) + (8.73 \times 10^3)$

(d) $(9.82 \times 10^7) - (6.31 \times 10^6)$

3 A computer performs 400 million calculations per second. Work out the number of calculations it performs in 2 hours. Give your answer in standard form.

4 The distance of the Earth from the Sun is $1.5 \times 10^8$ km. The distance of Uranus from the Sun is $2.8 \times 10^9$ km. Approximately how many times further from the Sun is Uranus than the Earth?

5 Astronomical distances are often measured in light years.

1 light year $= 9.46 \times 10^{12}$ km.

(a) The centre of the Milky Way is $2.6 \times 10^4$ light years from Earth. How many kilometres is this? Give your answer in standard form, correct to two significant figures.

(b) The Andromeda Galaxy is $1.4 \times 10^{19}$ km from Earth. Express this in light years. Give your answer in standard form, correct to two significant figures.

# 22

## statistics 2

**In this chapter you will learn:**

- how to find the mode, median, mean and range from lists and tables
- how to solve problems involving the mode, median, mean or range
- how to find an estimate for the mean from grouped frequency tables
- how to draw and use cumulative frequency graphs.

## 22.1 Averages

Chapter 06 dealt with the collection and organization of data. This chapter deals with the analysis of data. One way in which you can do this is to find an average, usually a single value, which is typical or representative of the set of data. The three averages most commonly used are the mode, the median and the arithmetic mean, usually just called the mean.

## 22.2 The mode

The **mode** is the item which occurs most often. In everyday speech, 'mode' means 'fashion'; you can think of the mode as the most fashionable or popular item. Thus, if the shoe sizes of seven men are 8, 6, 9, 7, 9, 7 and 9, the mode, or modal size, is 9, because it appears three times and no other size appears more than twice.

If the above information about shoe sizes is shown on a bar chart (see Figure 22.1), the mode is the shoe size with the highest bar.

To find the mode of longer lists of numbers, you might find it helpful to use tally marks and draw up a frequency table.

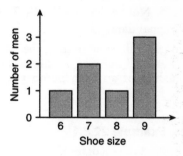

**Figure 22.1**

The mode is not always a number; it could, for example, be a colour. The mode has the advantages of being easy to obtain and being unaffected by extreme values but it is possible for data to have more than one mode or, if all the items are all different, no mode.

---

**Example 22.2.1**

Find the mode of
(a) 4, 6, 7, 5, 4, 7, 6, 8, 6, 8    (b) 2, 3, 5, 3, 1, 6, 5.

(a) 6 occurs 3 times. 4, 7 and 8 each occur twice. 5 occurs once, so the mode is 6.
(b) 3 and 5 each occur twice. 1, 2 and 6 each occur once. So the modes are 3 and 5. (The data is said to be **bi-modal**.)

---

## 22.3 The median

The **median** is the middle value, after the data have been arranged in order of size. For example, to find the median of 7, 3, 6, 4, 2, write the numbers in order of size (2, 3, 4, 6, 7) and find the middle number, which is 4. So the median is 4.

For five values, the median is the third value in the data, after they have been arranged in order of size. If there were seven values in the data, the median would be the fourth value and so on. In general, after you have put the values in order of size, you can find the position of the median by adding one to the number of values and dividing the result by 2. For example, if there are 17 values, $\frac{1}{2}(17+1)$, that is, the median is the ninth value.

A problem arises with the median of an even number of values, for example 7, 3, 8, 1, 9, 2. When you write the numbers in order of size (1, 2, 3, 7, 8, 9), there are two numbers in the middle, 3 and 7. The median is the number halfway between 3 and 7, that is, 5. Notice that $5 = \frac{1}{2}(3+7)$ and that 5 did not appear in the original data.

In general, after putting the values in order of size, the median of an odd number of values is simply the middle value and the median of an even number of values is halfway between the middle two values.

---

**Example 22.3.1**

Find the median of
(a) 2, 4, 5, 1, 4, 1, 3, 0, 2   (b) 13, 4, 12, 27, 3, 7.

(a) Write the numbers 0, 1, 1, 2, 2, 3, 4, 4, 5 in order of size. The middle number is 2, so the median is 2.
(b) Write the numbers 3, 4, 7, 12, 13, 27 in order of size. There are two numbers in the middle, 7 and 12. Adding 7 and 12 and dividing the result by 2 gives $\frac{1}{2}(7+12) = \frac{1}{2} \times 19 = 9\frac{1}{2}$, so the median is $9\frac{1}{2}$.

---

From (b) the median, like the mode, is not affected by extreme values.

## 22.4 The mean

The **mean** is the sum of the values divided by the number of values, that is, $\text{mean} = \dfrac{\text{sum of values}}{\text{number of values}}$.

For example, to find the mean of 9, 12, 2 and 5, add the numbers (28) and then divide 28 by 4, the number of values. Thus the mean is $\frac{9+12+2+5}{4} = \frac{28}{4} = 7$. The mean, therefore, need not be one of the numbers in the original data. Also, it need not be a whole number, even if all the values in the data are whole numbers.

---

**Example 22.4.1**

Ten boys were asked how many sisters each of them had. The numbers were 3, 2, 0, 2, 1, 4, 0, 1, 2, 1. Work out the mean number of sisters.

$$\text{Mean} = \frac{3+2+0+2+1+4+0+1+2+1}{10} = \frac{16}{10} = 1.6.$$

---

Of course, you cannot have 0.6 of a sister, but it is still correct to give this answer as 1.6.

The 'average' referred to in everyday speech is usually the mean. The term has already been used in this book; *average number of hours of sunshine* and *average monthly rainfall* occurred in Chapter 06 and *average speed* appeared in Chapter 08. Each of these is a mean value.

An advantage of the mean is that it makes use of all the data; a disadvantage is that it may be affected a great deal by extreme values.

---

**Example 22.4.2**

The rainfall, in millimetres, each day during a week is 0, 1, 2, 17, 1, 0, 0. Work out the mean daily rainfall and comment on your answer.

$$\text{Mean} = \frac{0+1+2+17+1+0+0}{7} = \frac{21}{7} = 3.$$

The mean rainfall is 3 mm.

The mean is not a very representative value, as, on six of the seven days, the rainfall is less than 3 mm. The median, 1 mm, might be a more appropriate representative value for certain purposes.

---

At the start of this section, you saw that, if the mean of four numbers is 7, their sum is 28, which is $7 \times 4$.

In general, sum of values = mean × number of values.

**Example 22.4.3**

The mean number of points scored by a rugby team in their first four matches was 17. After five matches, their mean was 19 points. How many points did they score in the fifth match?

Number of points scored in first four matches = $17 \times 4 = 68$.

Number of points scored in all five matches = $19 \times 5 = 95$.

Number of points scored in the fifth match = $95 - 68 = 27$.

## EXERCISE 22.1

1 Find the mode, median and mean of the following sets of numbers.
(a) 7, 7, 8, 11, 12        (b) 9, 5, 13, 5
(c) 9, 0, 8, 9, 4          (d) 7, 1, 6, 7, 6, 1, 7
(e) 12, 5, 7, 17, 7, 15    (f) 5, −1, −4, 0, 5

2 The number of children in each of ten families was 2, 3, 1, 3, 4, 2, 1, 3, 2, 3. Find the mode, median and mean.

3 The marks of five candidates in an examination were 76, 84, 72, 84 and 59. Find the mode, median and mean.

4 The numbers of goals scored by a soccer team in the first eight matches of the season were 3, 7, 1, 0, 2, 1, 5 and 1. Find the mode, median and mean.

5 The minimum daily temperatures, in °C, during a winter week were −1, −8, 0, −6, −1, −2 and −3. Find the mode, median and mean.

6 The lengths of six films are 2 h 4 min, 2 h 14 min, 1 h 43 min, 2 h 14 min, 2 h 29 min and 1 h 52 min. Find, in hours and minutes, the mode, median and mean.

7 The ages of four children are 3 years 7 months, 5 years 4 months, 6 years 8 months and 8 years 9 months. Find, in years and months, their mean age.

8 (a) A man drives 252 miles in 6 hours. Find his average speed.

(b) On the next day, he drives 105 miles at the same average speed. How long will this journey take him?

9 The mean of ten numbers is 8.3. Work out their sum.

10 The mean of 19, 37, 17, $x$, 29 and 15 is 24. Find the value of $x$.

11 The mode of five numbers is 1; the median is 5 and the mean is 4. Find the numbers.

12  The mean weight of seven people is 61 kg. Another person joins the group and the mean falls to 59 kg. Work out the weight of the eighth person.

13  The mean height of the 15 boys in a class is 162 cm and the mean height of the 10 girls is 152 cm. Work out the mean height of the 25 pupils.

14  A cyclist travels for 3 hours at 20 m.p.h. and then for 2 hours at 15 m.p.h. Work out her average speed for the whole journey.

# 22.5  The range

In addition to finding averages, it is useful to know how spread out data are. For example, if in an examination six boys scored 54, 55, 58, 62, 62, 63 and seven girls scored 32, 50, 54, 60, 62, 62, 93, the boys' mode (62), median (60) and mean (59) are the same as those of the girls but the spread, or dispersion, of each set of marks is quite different.

The **range** is one measure of the spread of data. The range is the difference between the highest value and the lowest value, that is,
    range = highest value − lowest value.

Thus, the range of the boys' marks is $63 - 54 = 9$ and the range of the girls' marks is $93 - 32 = 61$. Notice that, in mathematics, the range is a single value, even though, in everyday speech, you might describe the boys' range, for example, as 54 to 63.

The range has the advantage of being simple to work out but it is obviously affected by extreme values which are very different from other values in the data.

---

**Example 22.5.1**

In Mathsville, the number of hours of sunshine each day last week was 6, 9, 4, 3, 11, 6 and 7. Find the range.

The highest value is 11 and the lowest value is 3.

Highest value − lowest value = $11 - 3 = 8$, so the range is 8 hours.

---

## EXERCISE 22.2

1  Find the range of the following sets of numbers.
   (a) 2, 4, 7, 8, 9        (b) 24, 17, 42, 29, 37, 19
   (c) 3, −4, 0, −2, 1, −3

2 The range of a set of data is 9. The lowest value is 12. Work out the highest value.

3 In a science examination, six students scored 42, 76, 51, 64, 48, 59 and, in a history examination, they scored 86, 24, 93, 17, 41, 65. Find the range of their marks for each examination.

4 One day last winter, I measured the temperature every four hours. My readings, in °C, were −6, −3, 0, 3, 1, −2. Find the range.

5 The median of five numbers is 8; their mode is 3; their range is 9 and their mean is 7. Find the numbers.

## 22.6 Frequency tables

You can find averages and the range from a frequency table. For example, the table shows information about the number of goals scored by 20 soccer teams.

| Number of goals | 0 | 1 | 2 | 3 | 4 | 5 |
|---|---|---|---|---|---|---|
| Number of teams | 2 | 6 | 5 | 4 | 1 | 2 |

The mode, or modal number of goals, is 1, because more teams, 6, scored 1 goal than any other number of goals.

To find the median, you could first write down the numbers of goals from the table 0, 0, 1, 1, 1, 1, 1, 1, 2, 2, 2, 2, 2, 3, 3, 3, 3, 4, 5, 5.

The median lies halfway between the tenth and eleventh values (see Section 22.3) and, as these are both 2, the median is 2.

You can, however, find the median using the table without writing out a list. The first two teams scored no goals, so the second team scored no goals. The next six teams scored 1 goal, so the eighth team scored 1 goal. The next five teams scored 2 goals, so the thirteenth team scored 2 goals. The tenth and eleventh teams must each have scored 2 goals, so the median is 2.

The numbers 2, 8, 13 are called the running total or cumulative frequency. The next two are 17 (13 + 4) and 18 (17 + 1). The last one is 20, the total number of teams.

To work out the mean, you need to find the total number of goals and divide it by the number of teams, 20. Two teams scored no goals. Six teams each with 1 goal scored a total of 6 goals (1 × 6). five teams each with 2 goals scored a total of 10 goals (2 × 5) and so on.

Thus, the total number of goals
$$= (0 \times 2) + (1 \times 6) + (2 \times 5) + (3 \times 4) + (4 \times 1) + (5 \times 2)$$
$$= 0 + 6 + 10 + 12 + 4 + 10 = 42.$$

So the mean is $\frac{42}{20} = 2.1$.

The highest number of goals is 5 and the lowest is 0. The range is $5 - 0 = 5$, so the range is 5 goals.

### EXERCISE 22.3

Find the mode, median, mean and range from the following tables.

1  The table shows the number of goals scored by 25 hockey teams.

| Number of goals | 0 | 1 | 2 | 3 | 4 | 5 | 6 |
|---|---|---|---|---|---|---|---|
| Number of teams | 1 | 2 | 4 | 7 | 8 | 1 | 2 |

2  The table shows the number of days 30 people were absent from work last week.

| Number of days | 0 | 1 | 2 | 3 | 4 | 5 |
|---|---|---|---|---|---|---|
| Number of people | 9 | 8 | 4 | 5 | 3 | 1 |

3  The table shows the numbers of people living in houses in a street.

| Number of people | 1 | 2 | 3 | 4 | 5 | 6 | 7 |
|---|---|---|---|---|---|---|---|
| Number of houses | 3 | 6 | 5 | 2 | 3 | 0 | 1 |

4  According to the label on a box of screws, the 'average' contents are 15. John bought some of these boxes of screws and counted the number of screws in each box. The table shows his results.

| Number of screws | 13 | 14 | 15 | 16 | 17 |
|---|---|---|---|---|---|
| Number of boxes | 1 | 1 | 7 | 6 | 3 |

## 22.7  Grouped frequency tables

In Chapter 06, Section 6.9, you saw how to use class intervals to show data in grouped frequency tables. Even if you do not have the original data, you can use a grouped frequency table to obtain a variety of statistical information.

The marks scored by 50 students in a test were used to complete this grouped frequency table.

| Mark | 1–5 | 6–10 | 11–15 | 16–20 | 21–25 |
|------|-----|------|-------|-------|-------|
| Frequency | 7 | 9 | 12 | 17 | 5 |

You cannot find the mode from the table, but you can find which class interval has the highest frequency. This is called the **modal class interval**. So, in this example, the modal class interval is 16–20. The mode does not necessarily lie in the modal class interval.

Nor can you use the table to find the exact values of the mean and the median but you can use it to estimate them.

At one extreme, the seven students in the interval 1–5 could each have a mark of 1; at the other, they could each have a mark of 5. It is more likely, though, that their marks are distributed throughout the interval. To estimate the mean, you need only an approximate sum of the seven marks, so assume each of the seven students scored a mark in the middle of the interval. The halfway mark of the interval 1–5 is 3. So the total mark of the seven students is approximately $3 \times 7$, that is, 21. Continuing like this, you approximate the sum of all the students' marks.

$$\text{Sum} \approx 3 \times 7 + 8 \times 9 + 13 \times 12 + 18 \times 17 + 23 \times 5$$
$$= 21 + 72 + 156 + 306 + 115 = 670.$$

Finally, to obtain an estimate of the mean, divide the approximate sum, 670, by the number, 50, of students.

$$\text{Mean} \approx \frac{670}{50} = 13.4.$$

The median lies halfway between the 25th and 26th values. You can find the interval in which the median lies using an approach similar to that in Section 22.5.

The first seven students scored marks in the interval 1–5, so the seventh student scored a mark of 5 or less. The next nine students scored marks from 6–10, so the 16th student scored a mark of 10 or less. The next 12 students scored marks from 11–15, so the 28th student scored a mark of 15 or less. The 25th and 26th student must each have scored marks in the interval 11–15, so the median lies in this interval. In this case, the median lies in the middle interval, but this will not always be so.

The numbers 7, 16 and 28 are called **cumulative frequencies**. The next one is 45 (28 + 17) and the last one is 50, the total number of students. You can show the cumulative frequency in a table.

| Mark | up to 5 | up to 10 | up to 15 | up to 20 | up to 25 |
|------|---------|----------|----------|----------|----------|
| Cumulative frequency | 7 | 16 | 28 | 45 | 50 |

You can use the values in the table to draw a **cumulative frequency graph**. (See Figure 22.2). A cumulative frequency of 7 is plotted above a mark of 5, because you cannot be sure that all seven students have been included until you reach a mark of 5. A cumulative frequency of 16 is plotted above a mark of 10 and so on. You can join the points either with a smooth curve, as shown, or with a series of straight lines.

As this data is discrete and the marks are whole numbers, you plot the cumulative frequencies above 5, 10, 15, 20 and 25. Had the data been continuous, you should plot above 5.5, 10.5, 15.5, 20.5 and 25.5. However, in practice, this makes little difference in the results.

You can use the graph to obtain an estimate for the median by finding the mark which corresponds to a cumulative frequency of 25. (Strictly speaking, it is a cumulative frequency of $25\frac{1}{2}$ but 25 is near enough for an estimate.) The dotted lines with arrows in Figure 22.2 show how this method gives 14 as an estimate of the median.

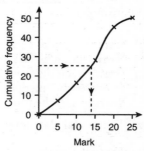

**Figure 22.2**

The median is the mark with a cumulative frequency 25, which is half of the total cumulative frequency, 50. The mark with a cumulative frequency which is one-quarter of the total cumulative frequency, $12\frac{1}{2}$, is called the **lower quartile** or **first quartile**.

From the graph, the lower quartile is 8. You can think of the median as the second quartile. Then the mark with a cumulative frequency of three-quarters of the total cumulative frequency, that is, $37\frac{1}{2}$, is the **upper quartile** or **third quartile**. From the graph, the upper quartile is 18.

The difference between the lower quartile and upper quartile is called the **interquartile range**, which is a measure of the spread of the marks. In this case, therefore, the interquartile range $= 18 - 8 = 10$.

You can also use the graph to find how many candidates scored less than a certain mark and, by subtraction from 50, how many scored more than that mark. For example, reading up to the curve from a mark of 19, you find that 42 students scored 19 marks or less so $50 - 42$, that is, eight students scored more than 19 marks.

In the next example, the class intervals for continuous data are written in a way you met in Chapter 06, Section 6.9.

**Example 22.7.1**

The grouped frequency table is about the heights of 80 women.

| Height ($h$ cm) | $140 \leq h < 150$ | $150 \leq h < 160$ |
| --- | --- | --- |
| Frequency | 4 | 29 |
| Height ($h$ cm) | $160 \leq h < 170$ | $170 \leq h < 180$ |
| Frequency | 38 | 9 |

(a) State the modal class interval.
(b) Work out an estimate of the mean.
(c) Draw a cumulative frequency graph and use it to estimate the median, the lower quartile, the upper quartile and the interquartile range.

(a) The modal class interval is $160 \leq h < 170$.

(b) Mean in (cm) $\approx \dfrac{145 \times 4 + 155 \times 29 + 165 \times 38 + 175 \times 9}{80}$

$= \dfrac{580 + 4495 + 6270 + 1575}{80} = \dfrac{12\,920}{80}$

$= 161.5$.

The mean is approximately 161.5 cm.

(c)

| Height ($h$ cm) | $140 \leq h < 150$ | $140 \leq h < 160$ |
| --- | --- | --- |
| Cumulative frequency | 4 | 33 |
| Height ($h$ cm) | $140 \leq h < 170$ | $140 \leq h < 180$ |
| Cumulative frequency | 71 | 80 |

Notice how the entries in the row marked 'heights' differ from the ones in the earlier table in this example. This is typical of tables which are constructed for drawing cumulative frequency graphs.

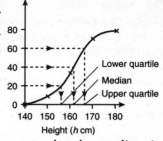

From the cumulative frequency graph, the median is 162 cm; the lower quartile is 157 cm and the upper quartile is 167 cm. The interquartile range is, therefore, 167 cm − 157 cm = 10 cm.

## EXERCISE 22.4

1 The table gives information about the number of words in 40 sentences in a book.

| Number of words | 1–5 | 6–10 | 11–15 | 16–20 |
|---|---|---|---|---|
| Frequency | 5 | 9 | 19 | 7 |

(a) State the modal class interval.
(b) Find an estimate of the mean.
(c) Draw a cumulative frequency graph to estimate the median, the lower and upper quartiles and the interquartile range.

2 The table gives information about the weights, in kilograms, of 100 men.

| Weight ($w$ kg) | Frequency |
|---|---|
| $60 \le w < 70$ | 8 |
| $70 \le w < 80$ | 14 |
| $80 \le w < 90$ | 23 |
| $90 \le w < 100$ | 39 |
| $100 \le w < 110$ | 16 |

(a) State the modal class interval.
(b) Find an estimate of the mean.
(c) From a cumulative frequency graph, estimate the median, the lower and upper quartiles and the interquartile range.
(d) How many men weighed more than 76 kg?

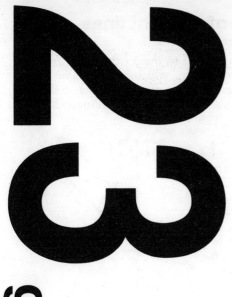

**23**

**graphs 2**

**In this chapter you will learn:**

- how to find equations of straight lines
- how to draw straight line graphs
- how to find and use the gradient of a straight line graph
- how to draw quadratic, cubic and reciprocal graphs

# 23.1 Equations of straight lines

You saw in Chapter 08, Section 8.3, how to find the equation of a straight line which is parallel to one of the axes. You can sometimes find the equation of a line which is not parallel to one of the axes by finding the coordinates of some points on the line and looking for a relationship between the $x$-coordinates and the $y$-coordinates.

For example, in Figure 23.1, the coordinates of some of the points on line (a) are (0,0), (1,1), (2,2) and (3,3). In general, the $y$-coordinate is equal to the $x$-coordinate, so the equation of line (a) is $y = x$ or $x = y$. The coordinates of some of the points on line (b) are $(-3,-1)$, $(-2,0)$, $(-1,1)$, (0,2), (1,3), (2,4) and (3,5). The relationship may be clearer if you show the coordinates in a table.

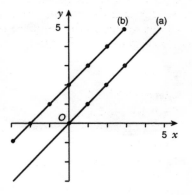

**Figure 23.1**

| $x$ | $-3$ | $-2$ | $-1$ | 0 | 1 | 2 | 3 |
|---|---|---|---|---|---|---|---|
| $y$ | $-1$ | 0 | 1 | 2 | 3 | 4 | 5 |

The $y$-coordinate is 2 more than the $x$-coordinate, so the equation of line (b) is $y = x + 2$. It is parallel to the line $y = x$ and is that line moved up 2 units. In a similar way, you could show that the line $y = x - 2$ is the line $y = x$ moved down 2 units.

The coordinates of some of the points on line (c) in Figure 23.2 are shown in the table.

| $x$ | $-1$ | 0 | 1 | 2 |
|---|---|---|---|---|
| $y$ | $-1$ | 0 | 2 | 4 |

The $y$-coordinate is twice the $x$-coordinate, so the equation of line (c) is $y = 2x$.

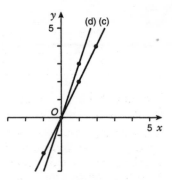

**Figure 23.2**

This table shows the coordinates of some of the points on line (d) in Figure 23.2.

| $x$ | $-1$ | 0 | 1 | 2 |
|---|---|---|---|---|
| $y$ | $-3$ | 0 | 3 | 6 |

The $y$-coordinate is 3 times the $x$-coordinate, so the equation of line (d) is $y = 3x$. In general, a line whose equation is $y = mx$ passes through the origin. The coefficient of $x$, 3 in the case of $y = 3x$, is related to the steepness of the line. The coefficient of $x$ may be negative or a fraction.

---

**Example 23.1.1**

Find the equation of each of the lines shown in the diagram.

The table shows the coordinates of some of the points on line (a).

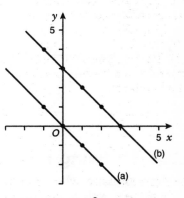

| $x$ | $-1$ | 0 | 1 | 2 |
|---|---|---|---|---|
| $y$ | 1 | 0 | $-1$ | $-2$ |

The $y$-coordinate is $-1$ times the $x$-coordinate, so the equation of line (a) is $y = -x$ or $x + y = 0$.

The table shows the coordinates of some of the points on line (b).

| $x$ | $-1$ | 0 | 1 | 2 | 3 |
|---|---|---|---|---|---|
| $y$ | 4 | 3 | 2 | 1 | 0 |

The sum of the $x$-coordinate and the $y$-coordinate is 3, so the equation of line (b) is $x + y = 3$.

Alternatively, you could say that line (b) is the line with equation $y = -x$ moved up 3 units, that is, the equation of line (b) is $y = -x + 3$, which you obtain when you make $y$ the subject of the equation $x + y = 3$.

Notice that in both $y = -x$ and $y = -x + 3$, the coefficient of $x$ is negative $(-1)$ and that the lines are sloping in the opposite direction to those on the previous page, which have a positive coefficient of $x$ in their equations.

## EXERCISE 23.1

1 Find the equation of each of the lines shown in the diagrams.

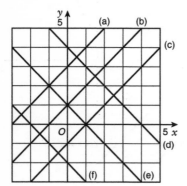 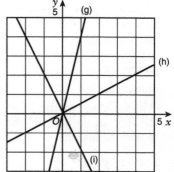

## 23.2 Drawing straight line graphs

To draw a graph when you are told its equation, you have to substitute values of $x$ into the equation and find the corresponding $y$-values. This gives you the coordinates of points on the graph.

For example, when substituting the value $x = 3$ into the equation $y = 2x + 1$, $y = 2 \times 3 + 1 = 7$. Thus $(3,7)$ is a point on the graph of $y = 2x + 1$. The coordinates are often shown in a table of values. This table shows the $x$- and $y$-coordinates of points on the graph of $y = 2x + 1$ for values of $x$ between $x = -2$ and $x = 4$.

| $x$ | $-2$ | $-1$ | 0 | 1 | 2 | 3 | 4 |
|-----|------|------|---|---|---|---|---|
| $y$ | $-3$ | $-1$ | 1 | 3 | 5 | 7 | 9 |

You can then plot the points and join them with a straight line (see Figure 23.3).

The line cuts the $y$-axis at $(0,1)$. This point is called its **intercept** on the $y$-axis.

In general, the graph of an equation of the form $y = mx + c$ is a straight line with $(0, c)$ as its intercept on the $y$-axis. For example, when $m = 3$ and $c = -2$, the equation is $y = 3x - 2$. Thus, the graph of $y = 3x - 2$ is a straight line and its intercept on the $y$-axis is $(0, -2)$.

If you are not asked to complete a table of values like the one above, you need plot only two points on a line to be able to draw it, but a third point is a useful check.

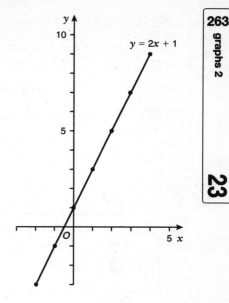

**Figure 23.3**

---

**Example 23.2.1**

(a) Draw the line whose equation is $y = -2x + 5$ for values of $x$ between $x = -2$ and $x = 4$.

(b) Write down the coordinates of its intercept on the $y$-axis.

(a) Choose three suitable $x$ values, say $x = 0$, $x = 1$ and $x = 2$ and substitute these values into the equation $y = -2x + 5$.

When $x = 0$, $y = 0 + 5 = 5$; when $x = 1$, $y = -2 + 5 = 3$; when $x = 2$, $y = -4 + 5 = 1$.

With practice, you should be able to do this working mentally and put the $y$-value obtained using each $x$-value straight into a table of values.

| $x$ | 0 | 1 | 2 |
|---|---|---|---|
| $y$ | 5 | 3 | 1 |

Plot the points with coordinates (0,5), (1,3) and (2,1). Join them with a straight line and extend it in both directions to cover the range of x-values specified, in this case, $x = -2$ to $x = 4$. (See diagram.)

(b) The intercept on the y-axis is (0,5).

Although equations of lines are often written in the form $y = mx + c$, you can use algebra to express them in other ways. The equation $y = -2x + 5$, for example, may be rearranged and written as $y = 5 - 2x$ or $2x + y = 5$ or any other equivalent form.

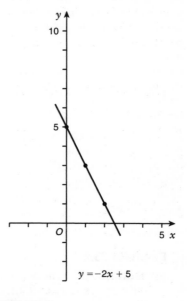

$y = -2x + 5$

## EXERCISE 23.2

1 Complete the table for the equation $y = 3x - 2$ and draw its graph.

| x | −2 | −1 | 0 | 1 | 2 | 3 | 4 |
|---|----|----|---|---|---|---|---|
| y |    | −5 |   |   |   | 7 |   |

2 Complete the table for the equation $y = \frac{1}{2}x + 1$ and draw its graph.

| x | −6 | −4 | −2 | 0 | 2 | 4 | 6 |
|---|----|----|----|---|---|---|---|
| y | −2 |    |    |   |   | 3 |   |

3  Complete the table for the equation $y = -2x - 3$ and draw its graph.

| $x$ | -2 | -1 | 0 | 1 | 2 | 3 | 4 |
|---|---|---|---|---|---|---|---|
| $y$ | | -1 | | | | | |

4  Draw lines with the following equations for values of $x$ between $x = -4$ and $x = 4$. For each line, write down the coordinates of its intercept on the $y$-axis.
(a) $y = 4x + 3$    (b) $y = -3x + 4$    (c) $y = \frac{1}{3}x - 2$
(d) $y = 3 - \frac{1}{2}x$    (e) $x + y = 2$    (f) $3x + y = 6$

## 23.3  The gradient of a straight line

Figure 23.4 shows the lines with equations $y = 2x + 3$, $y = 2x + 1$ and $y = 2x - 1$. You saw in Section 23.1 that the coefficient of $x$ affects the steepness of a line. In this case, the coefficient, 2, of $x$ is the same in all three equations and the lines are parallel.

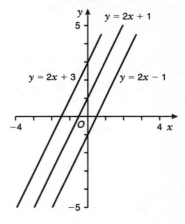

You saw in Section 23.2 that, using the $+3$, $+1$ and $-1$ in the equations, you can write down the intercepts on the $y$-axis of the lines as $(0,3)$, $(0,1)$ and $(0,-1)$ respectively.

**Figure 23.4**

Thus, you can write down the equation of any line parallel to these lines, if you know its intercept on the $y$-axis. For example, if its intercept on the $y$-axis is $(0, -5)$, its equation is $y = 2x - 5$.

---

**Example 23.3.1**

A line passes through the point with coordinates $(2, 10)$ and is parallel to the line whose equation is $y = 3x - 1$. Find the equation of the line and its intercept on the $y$-axis.

The equation of the line is $y = 3x + c$. When $x = 2$, $y = 10$, so $10 = 3 \times 2 + c$, that is $c = 4$. Thus, the equation is $y = 3x + 4$ and the intercept is $(0,4)$.

---

To find a more precise interpretation than 'steepness' of the coefficient, $m$, of $x$ in the equation $y = mx + c$, consider the line with equation $y = 2x + 1$, shown in Figure 23.5.

$A, B$ and $C$ are points on the line.

The value of $\dfrac{BE}{AE} = \dfrac{2}{1} = 2$.

The value of $\dfrac{CD}{BD} = \dfrac{4}{2} = 2$.

In fact, for any two points on the line, $\dfrac{\text{the } y\text{-step}}{\text{the } x\text{-step}}$ is always 2. This is the **gradient** of the line. (It is also the tangent of the angle the line makes with the horizontal.)

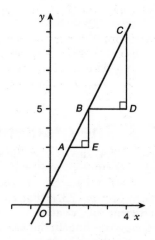

**Figure 23.5**

In the equation $y = mx + c$, $m$ is the gradient of the line and its intercept on the $y$-axis is $(0, c)$. For example, the line whose equation is $y = 6x - 5$ has a gradient of 6 and its intercept on the $y$-axis is $(0, -5)$.

In a similar way, the line with equation $y = -2x + 5$ (see Example 23.2.1) has a gradient of $-2$ and its intercept is $(0, 5)$. You can see from its graph the direction of the slope of a line with a negative gradient. A horizontal line has a gradient of zero.

To find the gradient of a line joining a pair of points, start by drawing a sketch. It need not be accurate but the points must be in the correct relative positions. This will enable you to decide whether the gradient of the line joining them is positive or negative.

In general, to find the gradient of the line joining the points with coordinates $(a, b)$ and $(c, d)$, you should use either

$$\text{gradient} = \frac{d - b}{c - a} \quad \text{or} \quad \text{gradient} = \frac{b - d}{a - c}.$$

**Example 23.3.2**

Find the gradient of the line joining the points with coordinates

(a) (5,3) and (2,1), (b) (3,8) and (5,2).

(a) From the sketch, which is not to scale, the gradient is positive. The $y$-step is $3 - 1 = 2$ and the $x$-step is $5 - 2 = 3$.

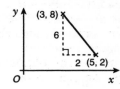

$$\text{Gradient} = \frac{\text{the } y\text{-step}}{\text{the } x\text{-step}} = \tfrac{2}{3}.$$

(b) When you sketch the points, you find that the line joining them slopes in the other way and the gradient is therefore negative.

The $y$-step is $8 - 2 = 6$ and the $x$-step is $3 - 5 = -2$.

$$\text{Gradient} = \frac{\text{the } y\text{-step}}{\text{the } x\text{-step}}$$

$$= \frac{6}{-2} = -3.$$

**Example 23.3.3**

Find the equation of the line joining the points (3,8) and (5,2).

From Example 23.3.2(b), the gradient is $-3$.

Using $y = mx + c$, the equation of the line is $y = -3x + c$.

When $x = 3$, $y = 8$ so $8 = -3 \times 3 + c$, that is, $c = 17$.

Therefore the equation of the line is $y = -3x + 17$.

As a check, notice that the values $x = 5$, $y = 2$ satisfy this equation.

Sometimes you may have to rearrange the equation of the line in the form $y = mx + c$, that is, make $y$ the subject.

---

**Example 23.3.4**

Find (a) the gradient and (b) the intercept on the $y$-axis of the line whose equation is $3x - 2y = 12$.

Rearrange the equation as $\quad 2y = 3x - 12$.

Make $y$ the subject $\qquad\qquad y = \frac{3}{2}x - 6$

The gradient is $\frac{3}{2}$ and the intercept on the $y$-axis is $(0, -6)$.

---

## EXERCISE 23.3

1 The equation of a line is $y = 7x - 4$. Write down
   (a) the gradient of the line,   (b) its intercept on the $y$-axis.

2 The gradient of a line is $-5$ and its intercept on the $y$-axis is $(0,4)$. Write down the equation of the line.

3 A line passes through the point with coordinates $(4,13)$ and is parallel to the line whose equation is $y = 2x - 3$. Find the equation of the line and its intercept on the $y$-axis.

4 Find the gradient of the line joining the points with coordinates
   (a) $(3,2)$, $(5,10)$   (b) $(-3,5)$, $(5,1)$   (c) $(-2,4)$, $(5,4)$.

5 Find the equation of the line joining the points with coordinates
   (a) $(0,3)$, $(3,15)$   (b) $(4,5)$, $(10,8)$   (c) $(1,7)$, $(4,-8)$.

6 Find the gradient and the intercept on the $y$-axis of the line
   (a) $2x + y = 3$,   (b) $2x - 5y = 15$,   (c) $3x + 4y = 24$.

# 23.4 Curved graphs

Not all graphs are straight lines. The graph of $y = x^2$, for example, is curved. Here is a table of values for the equation $y = x^2$.

| $x$ | $-3$ | $-2$ | $-1$ | 0 | 1 | 2 | 3 |
|-----|------|------|------|---|---|---|---|
| $y$ | 9 | 4 | 1 | 0 | 1 | 4 | 9 |

Figure 23.6 shows the points from the table and the graph of $y = x^2$, formed by joining the points with a curve, called a **parabola**.

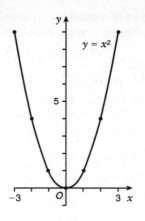

$y = x^2$

**Figure 23.6**

This is the simplest **quadratic** graph. The word 'quadratic' comes from the Latin *quadratus,* meaning 'square'. In the equation of a quadratic graph, the highest power of $x$ is $x^2$. Thus, $y = x^2 + 3$, $y = 3x^2 - 2x + 1$ and $y = 4x^2 - 3x$ are all equations of quadratic graphs.

Note that although Figure 23.6 is drawn with the same scale on both axes, it can be difficult to see all the features of the graph. It is usual to use different scales on the axes if necessary.

---

### Example 23.4.1

Draw the graph of $y = x^2 - 2x - 1$ for values of $x$ from $-2$ to $4$.

When $x = -2$, $y = (-2)^2 - 2 \times (-2) - 1 = 4 + 4 - 1 = 7$. Making similar calculations, you can complete the table of values.

| $x$ | $-2$ | $-1$ | $0$ | $1$ | $2$ | $3$ | $4$ |
|-----|------|------|-----|-----|-----|-----|-----|
| $y$ | $7$ | $2$ | $-1$ | $-2$ | $-1$ | $2$ | $7$ |

The diagram right shows the graph of $y = x^2 - 2x - 1$.

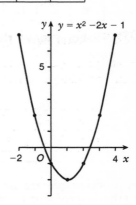

$y = x^2 - 2x - 1$

Graphs of equations in which the highest power of $x$ is $x^3$ are called **cubic** graphs. For example, $y = 2x^3 - 9$ and $y = x^3 + 4x^2 - 2x + 7$ are equations of cubic graphs. The simplest cubic graph is $y = x^3$.

Here is a table of values for the equation $y = x^3$.

| $x$ | $-3$ | $-2$ | $-1$ | 0 | 1 | 2 | 3 |
|---|---|---|---|---|---|---|---|
| $y$ | $-27$ | $-8$ | $-1$ | 0 | 1 | 8 | 27 |

Figure 23.7 shows the graph of $y = x^3$.

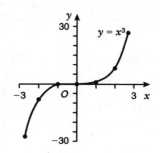

**Figure 23.7**

**Example 23.4.2**

Draw the graph of $y = x^3 + x^2 - 8x$ for values of $x$ from $-3$ to 3. First, draw up a table of values.

| $x$ | $-3$ | $-2$ | $-1$ | 0 | 1 | 2 | 3 |
|---|---|---|---|---|---|---|---|
| $y$ | 6 | 12 | 8 | 0 | $-6$ | $-4$ | 12 |

The diagram shows the graph of $y = x^3 + x^2 - 8x$.

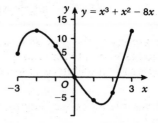

**Reciprocal graphs** are the final type included in this chapter. You find the reciprocal of a number by dividing it into 1. For example, the reciprocal of 2 is $\frac{1}{2}$. So a reciprocal graph has an equation such as $y = \frac{2}{x}$ or $y = 3 - \frac{4}{x}$. The simplest reciprocal graph is $y = \frac{1}{x}$.

Here is a table of values for $y = \frac{1}{x}$. As division by 0 is undefined, $y$ is undefined when $x = 0$ and so values of $x$ *near* 0 are used in the table.

| $x$ | $-3$ | $-2$ | $-1$ | $-0.5$ | $-0.2$ | $0.2$ | $0.5$ | $1$ | $2$ | $3$ |
|---|---|---|---|---|---|---|---|---|---|---|
| $y$ | $-0.3$ | $-0.5$ | $-1$ | $-2$ | $-5$ | $5$ | $2$ | $1$ | $0.5$ | $0.3$ |

Figure 23.8 shows the graph of $y = \frac{1}{x}$. It is unlike the graphs you have met so far in this chapter in that there is a break in the graph at $x = 0$. The $x$- and $y$-axes have the property that the graph gets infinitely close to them, but never actually touches them.

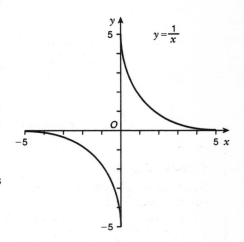

**Figure 23.8**

## EXERCISE 23.4

The graphs in the text were drawn with grid lines because the features stand out more clearly that way. To answer the following questions you will need to use graph paper with a 2 mm grid.

1  Draw the following quadratic graphs for values of $x$ from $-4$ to 4.
  (a) $y = x^2 + 3$    (b) $y = 5x^2$    (c) $y = \frac{1}{2}x^2$
  (d) $y = -x^2$       (e) $y = -2x^2$   (f) $y = x^2 - 2x$
  (g) $y = (x - 1)^2$  (h) $y = x^2 - 4x - 1$  (i) $y = 2x^2 + 3x - 5$

2 Draw the following cubic graphs. Take values of $x$ from $-3$ to $3$.

   (a) $y = x^3 + 8$   (b) $y = 2x^3$         (c) $y = \frac{1}{2}x^3$

   (d) $y = -x^3$     (e) $y = -2x^3$     (f) $y = (x - 1)^3$

   (g) $y = x^3 - x^2$  (h) $y = x^3 - x^2 - 2x$  (i) $y = x^3 + 3x - 2$

3 Draw the following reciprocal graphs. Use the same $x$ values as those used for the graph of $y = \dfrac{1}{x}$.

   (a) $y = \dfrac{2}{x}$   (b) $y = -\dfrac{1}{x}$   (c) $y = -\dfrac{2}{x}$   (d) $y = \dfrac{1}{x} + 5$

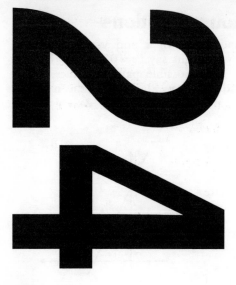

**24**

**equations 2**

**In this chapter you will learn:**
- how to solve simultaneous equations
- how to expand and simplify products which give quadratic expressions
- how to factorize quadratic expressions
- how to solve quadratic equations by factorization
- how to solve cubic equations by trial and improvement.

## 24.1 Simultaneous equations

If you were asked to find values of $x$ and $y$ which satisfy the equation $2x + y = 4$, you might say '$x = 0$, $y = 4$'. This is correct but it is not the only possible pair of values. Two more are $x = 1$, $y = 2$ and $x = 2$, $y = 0$. In fact, the number of pairs of values of $x$ and $y$ which satisfy $2x + y = 4$ is infinite.

From Chapter 23, you know that the graph of $2x + y = 4$ is a straight line (See Figure 24.1). The $x$-coordinate and $y$-coordinate of every point on this line represent a solution to the equation $2x + y = 4$.

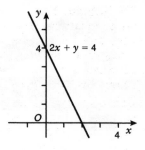

**Figure 24.1**

Similarly, the equation $x + y = 3$ has an infinite number of solutions. For example, $x = 3$, $y = 0$ and $x = 1$, $y = 2$ satisfy $x + y = 3$.

There is, however, only one pair of values which satisfy both $2x + y = 4$ and $x + y = 3$ at the same time, that is, simultaneously. This pair is $x = 1$, $y = 2$. The $x$- and $y$-coordinates of the point of intersection of the lines whose equations are $2x + y = 4$ and $x + y = 3$. These equations are called **simultaneous equations**. Their solution is $x = 1$, $y = 2$. To check, substitute this pair of values into the equations. In $2x + y = 4$, $2 \times 1 + 2 = 4$ and, in $x + y = 3$, $1 + 2 = 3$.

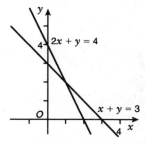

**Figure 24.2**

### Example 24.1.1

Solve the simultaneous equations $3x + y = 7$ and $2x - y = 8$.
Draw up tables of values of $3x + y = 7$ and $2x - y = 8$.

| $x$ | 0 | 1 | 2 |
|-----|---|---|---|
| $y$ | 7 | 4 | 1 |

| $x$ | 4 | 5 | 6 |
|-----|---|---|---|
| $y$ | 0 | 2 | 4 |

The diagram shows the lines with equations $3x + y = 7$ and $2x - y = 8$. The lines intersect at the point with coordinates $(3, -2)$, so the solution is $x = 3$, $y = -2$.

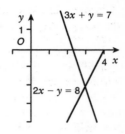

Checking in the original equations, $3 \times 3 + (-2) = 7$ and $2 \times 3 - (-2) = 8$.

## EXERCISE 24.1

1 Find three pairs of values of $x$ and $y$ which satisfy these equations.
   (a) $x + y = 5$     (b) $x - y = 3$     (c) $2x + 3y = 6$

2 Which of these pairs of values satisfy the equation $2x + 5y = 20$?
   (a) $x = 5, y = 2$     (b) $x = 4, y = 0$     (c) $x = -5, y = 6$

3 Which of these pairs of values satisfies the simultaneous equations $3x + 4y = 8$, $2x - 3y = 11$?
   (a) $x = 0, y = 2$     (b) $x = 4, y = -1$     (c) $x = 8, y = -4$

4 Draw appropriate straight lines to solve these simultaneous equations. Check your answers.
   (a) $x + y = 4$, $x - y = 2$          (b) $x + y = 5$, $2x + y = 7$
   (c) $x + 2y = 8$, $x - y = 2$         (d) $3x + 2y = 6$, $x - 2y = 10$
   (e) $x + y = 2$, $2x + 3y = 8$        (f) $x - y = 1$, $3x - 2y = 1$

5 Try to solve the simultaneous equations $x + y = 3$, $x + y = 5$ by drawing appropriate straight lines. What happens?

## 24.2 Algebraic methods

Graphical methods are time-consuming and may give only approximate solutions. Using algebra, you can **eliminate** one of the unknowns so that you are left with an equation in only one unknown.

Sometimes you can achieve this simply by adding the equations. For example, to solve the simultaneous equations $2x + 3y = 11$, $5x - 3y = 17$, you can add the two left-hand sides, $2x + 3y + (5x - 3y) = 2x + 3y + 5x - 3y = 7x$ and then add the two right-hand sides, $11 + 17 = 28$. Thus, $7x = 28$, so $x = 4$.

To find the value of $y$, substitute $x = 4$ into either of the original equations. If you substitute $x = 4$ into $2x + 3y = 11$, $2 \times 4 + 3y = 11$, that is, $3y + 8 = 11$, giving $y = 1$. The solution is $x = 4$, $y = 1$.

Checking by substituting $x = 4$, $y = 1$ into the equation $5x - 3y = 17$ gives $5 \times 4 - 3 \times 1 = 20 - 3 = 17$, so the solution is correct.

You have seen that, if the coefficients of an unknown differ only in sign, you can eliminate that unknown by adding two simultaneous equations. If, however, the coefficients of an unknown are equal, you can eliminate that unknown by subtracting one of the equations from the other.

---

**Example 24.2.1**

Solve the simultaneous equations $4x + 5y = 2$, $4x - 3y = 18$. Label the equations $A$ and $B$, and write them under each other.

$$4x + 5y = 2 \quad A$$
$$4x - 3y = 18 \quad B$$

Then you can use these labels to explain your method. Subtracting equation $B$ from equation $A$ gives

$$4x + 5y - (4x - 3y) = 2 - 18$$
$$4x + 5y - 4x + 3y = -16$$
$$8y = -16$$
$$y = -2.$$

Substituting $y = -2$ in $A$ gives $4x - 10 = 2$ so $x = 3$. The solution is $x = 3$, $y = -2$.

Check in $B$. $4 \times 3 - (3 \times (-2)) = 12 - (-6) = 18$.

In this example, you could subtract equation $A$ from equation $B$. The disadvantage of this is that it gives you a negative coefficient of $y$.

If you cannot eliminate one of the unknowns just by adding or subtracting the equations, you will have to multiply one or both of the equations by an appropriate number and then add or subtract.

---

**Example 24.2.2**

Solve the simultaneous equations $4x + 3y = 10$, $5x + 6y = 8$.

$$4x + 3y = 10 \quad A$$
$$5x + 6y = 8 \quad B$$

Multiply both sides of $A$ by 2 to make the coefficients of $y$ equal.

Label the new equation $C$.

$$8x + 6y = 20 \quad C$$

$C - B$ gives $3x = 12$, that is, $x = 4$.
Substitute $x = 4$ in $A$ giving $4 \times 4 + 3y = 10$, or $y = -2$.
The solution is $x = 4$, $y = -2$.
Check in $B$. $5 \times 4 + 6 \times (-2) = 20 - 12 = 8$.

---

Notice that you should make the check in one of the original two equations. You know these are correct but there could be an error in one of those you have obtained by multiplying.

In the next example, you have to multiply both equations. It also shows the usual way in which the steps in the working are described.

---

**Example 24.2.3**

Solve the simultaneous equations $5x + 2y = 7$, $4x - 3y = 24$.

$$5x + 2y = 7 \quad A$$
$$4x - 3y = 24 \quad B$$

Multiplying $A$ by 3 and $B$ by 2 and re-labelling, gives

$$15x + 6y = 21 \quad C$$
$$8x - 6y = 48 \quad D$$

Adding $C$ and $D$ gives $15x + 6y + (8x - 6y) = 21 + 48$, which is $23x = 69$, that is, $x = 3$.

Substituting $x = 3$ in $A$ gives $5 \times 3 + 2y = 7$, or $y = -4$.

The solution is $x = 3$, $y = -4$.

Check in $B$. $4 \times 3 - (3 \times (-4)) = 12 - (-12) = 12 + 12 = 24$.

In the above example, you could have started by multiplying $A$ by 4 and $B$ by 5. Then, by subtraction, you could have eliminated $x$.

You can use the elimination methods to solve problems in which the information can be expressed as a pair of simultaneous equations.

### Example 24.2.4

The sum of two numbers is 59 and their difference is 15. Find the numbers.

Let $x$ and $y$ be the numbers. Then you can write the given information as two simultaneous equations in the form

$$x + y = 59 \qquad A$$
$$x - y = 15 \qquad B$$

Adding $A$ and $B$ gives $2x = 74$, or $x = 37$.

Substituting in $A$, $37 + y = 59$, giving $y = 22$.

The numbers are 37 and 22.

Check in original problem. $37 + 22 = 59$ and $37 - 22 = 15$.

### EXERCISE 24.2

1 Solve the following pairs of simultaneous equations.
   (a) $3x + y = 17$   (b) $2x + 3y = 7$   (c) $5x + 4y = 11$
       $2x - y = 8$        $2x + 7y = 11$       $3x - 4y = 13$

2 Solve the following pairs of simultaneous equations.
   (a) $3x - 2y = 1$   (b) $5x + 3y = 7$   (c) $2x + 3y = 12$
       $4x + y = 5$        $7x + 9y = 5$       $3x - 4y = 1$
   (d) $3x + 5y = 6$   (e) $4x - 3y = 4$   (f) $8x - 3y = 16$
       $2x + 3y = 5$        $6x + 5y = 25$      $6x - 5y = 23$

3 The sum of two numbers is 74 and their difference is 24. Find the numbers.

4 Cinema tickets for two adults and a child cost £13. The cost for four adults and three children is £29. Find the cost of a child's ticket.

5 The sum of two numbers is 122. The difference between twice one number and three times the other is 49. Find the numbers.

6 The equation of a straight line is $y = mx + c$. When $x = 4$, $y = -1$ and when $x = 10$, $y = 2$. Find the values of $m$ and $c$.

7 The points with coordinates $(4, -1)$ and $(10, 2)$ lie on the line whose equation $ax + by = 6$. Find the values of $a$ and $b$.

## 24.3 Quadratic expressions

A quadratic expression is one in which the highest power of $x$ is $x^2$. Examples of quadratic expressions are $x^2 + 2x - 3$, $3x^2 + 5x$ and $4x^2 - 1$. Some quadratic expressions can be written as a product of factors, such as $2x(x - 3)$ and $(x + 4)(x + 3)$. You saw in Chapter 11, Section 11.6, how to expand expressions like $2x(x - 3)$ and the method used there can be extended to expand expressions like $(x + 4)(x + 3)$.

It can be helpful to think of $(x + 4)(x + 3)$ as $x$ 'lots' of $x + 3$ and 4 'lots' of $x + 3$. Then

$$\begin{aligned}
(x + 4)(x + 3) &= x(x + 3) + 4(x + 3) \\
&= x^2 + 3x + 4x + 12 \\
&= x^2 + 7x + 12.
\end{aligned}$$

Figure 24.3 illustrates the result $(x + 4)(x + 3) = x^2 + 7x + 12$.

You can think of the product $(x + 4)$ $(x + 3)$, as the area of a rectangle of length $x + 4$ and width $x + 3$.

The total area $(x + 4)(x + 3)$ is the sum of the areas of the four smaller rectangles.

Notice that, before simplification, there are four terms in the expansion.

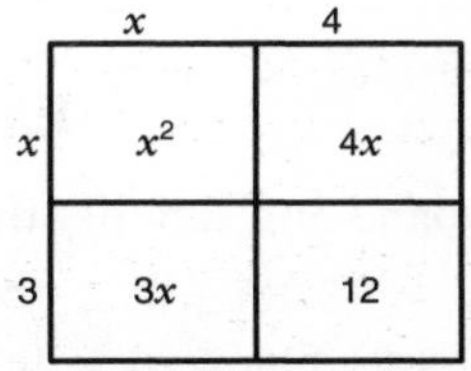

**Figure 24.3**

### Example 24.3.1

Expand and simplify $(x + 2)(x + 5)$.

$$\begin{aligned}(x + 2)(x + 5) &= x(x + 5) + 2(x + 5)\\&= x^2 + 5x + 2x + 10\\&= x^2 + 7x + 10.\end{aligned}$$

With practice, you should be able to omit the first line of working.

### Example 24.3.2

Expand and simplify the products   (a) $(x + 7)(x + 2)$,
(b) $(x - 5)(x + 3)$,   (c) $(x - 1)^2$,   (d) $(x + 4)(x - 4)$,
(e) $(4x - 5)(2x + 3)$,   (f) $(3x + 1)(5 - x)$.

(a) $\begin{aligned}[t](x + 7)(x + 2) &= x^2 + 2x + 7x + 14\\&= x^2 + 9x + 14.\end{aligned}$

(b) $\begin{aligned}[t](x - 5)(x + 3) &= x^2 + 3x - 5x - 15\\&= x^2 - 2x - 15.\end{aligned}$

(c) $\begin{aligned}[t](x - 1)^2 &= (x - 1)(x - 1)\\&= x^2 - x - x + 1\\&= x^2 - 2x + 1.\end{aligned}$

(d) $\begin{aligned}[t](x + 4)(x - 4) &= x^2 - 4x + 4x - 16\\&= x^2 - 16.\end{aligned}$

This is an example of the **difference of two squares**.

(e) $\begin{aligned}[t](4x - 5)(2x + 3) &= 8x^2 + 12x - 10x - 15\\&= 8x^2 + 2x - 15.\end{aligned}$

(f) $\begin{aligned}[t](3x + 1)(5 - x) &= 15x - 3x^2 + 5 - x\\&= 5 + 14x - 3x^2.\end{aligned}$

### EXERCISE 24.3

1 Expand and simplify these products.

  (a) $(x + 3)(x + 2)$     (b) $(x + 4)(x - 5)$     (c) $(x - 5)(x + 1)$
  (d) $(x - 6)(x - 3)$     (e) $(x + 5)^2$     (f) $(x - 2)^2$
  (g) $(x + 7)(x - 7)$     (h) $(x - 4)(x + 4)$     (i) $(x + 3)(6 - x)$
  (j) $(5 - x)(x + 1)$     (k) $(2x + 5)(3x + 4)$ (l) $(5x + 1)(4x - 3)$
  (m) $(6x - 1)(2x - 7)$ (n) $(3x + 2)^2$     (o) $(2x - 5)^2$
  (p) $(9x + 4)(9x - 4)$ (q) $(3x - 7)(3x + 7)$ (r) $(5x + 2)(6 - x)$
  (s) $(4x + 3)(2 - 3x)$ (t) $(5 - 3x)^2$     (u) $(ax - b)(ax + b)$

# 24.4 Factorizing quadratic expressions

You saw in Chapter 11, Section 11.7, how to factorize quadratic expressions with a common factor. For example, $x^2 - 5x = x(x - 5)$.

A different approach is, however, needed to factorize other types of quadratic expression.

To factorize $x^2 + 3x + 2$, you would seek two bracketed expressions with a product of $x^2 + 3x + 2$.

The first term in each bracket must be $x$.

As the number term, called the **constant term,** is positive, $+2$, the signs in the brackets must be the same, either both $+$ or both $-$.

The coefficient of $x$ is positive, so the signs in the brackets must both be $+$.

The factors are $(x + \ldots)(x + \ldots)$, where the product of the missing numbers must be 2.

The only possible pair of numbers is 1 and 2, so the factors are $(x + 1)(x + 2)$.

You should check that the answer is correct by finding the product of the factors, $(x + 1)(x + 2) = x^2 + 2x + x + 2 = x^2 + 3x + 2$.

---

**Example 24.4.1**

Factorize $x^2 - x - 12$.

As the sign of the constant term is negative, the signs inside the brackets must be different.

The factors are $(x + \ldots)\,(x - \ldots)$, where the product of the missing numbers is 12.

The possible pairs of numbers are 1 and 12, 2 and 6, and 3 and 4.

The factors must be one of these pairs.

$(x + 1)(x - 12) \quad (x + 2)(x - 6) \quad (x + 3)(x - 4)$
$(x - 1)(x + 12) \quad (x - 2)(x + 6) \quad (x - 3)(x + 4)$

By a process of trying these products one by one, you find that $x^2 - x - 12 = (x + 3)(x - 4)$.

Check: $(x + 3)(x - 4) = x^2 - 4x + 3x - 12 = x^2 - x - 12$.

---

You should have noticed that the process of factorization is a matter of following up the clues which are there. There is no sure-fire process for factorizing in this way.

### Example 24.4.2

Factorize $x^2 - 36$.

This is a difference of two squares. Example 24.3.2(d) suggests that $x^2 - 36 = (x + 6)(x - 6)$.

Check: $(x + 6)(x - 6) = x^2 - 6x + 6x - 36 = x^2 - 36$.

### EXERCISE 24.4

1 One factor of a quadratic expression is given. Find the other factor.

(a) $x^2 + 6x + 8$, $(x + 2)$      (b) $x^2 - 2x - 8$, $(x - 4)$
(c) $x^2 + 2x - 8$, $(x - 2)$      (d) $x^2 - 6x + 8$, $(x - 2)$
(e) $x^2 + 9x + 8$, $(x + 1)$      (f) $x^2 - 7x - 8$, $(x + 1)$
(g) $x^2 - 7x + 12$, $(x - 3)$      (h) $x^2 - 4$, $(x + 2)$
(i) $x^2 - 100$, $(x - 10)$      (j) $x^2 + 2x$, $(x)$

2 Factorize the following quadratic expressions.

(a) $x^2 - 2x - 15$      (b) $x^2 + 9x + 14$
(c) $x^2 - 24x - 25$      (d) $x^2 - 10x - 24$
(e) $x^2 - 49$      (f) $x^2 + 8x$
(g) $x^2 + 10x - 56$      (h) $x^2 - 1$
(i) $x^2 - 8x - 20$      (j) $x^2 - 9x$

## 24.5 Quadratic equations

A quadratic equation is one in which the highest power of $x$ is $x^2$, for example, $x^2 - 4 = 0$. If you were asked to solve this equation, you might say '$x = 2$'. This certainly satisfies $x^2 - 4 = 0$, as $2^2 - 4 = 0$. But there is a second solution, $x = -2$, because $(-2)^2 - 4 = 0$ also.

Figure 24.4 is the graph of $y = x^2 - 4$. It crosses the $x$-axis, that is, $y = 0$, at points whose $x$-coordinates are 2 and $-2$, the solutions of $x^2 - 4 = 0$. This suggests that you can solve any quadratic equation by drawing the related parabola and finding the $x$-coordinate of each of the points where it cuts the $x$-axis. There are usually two solutions.

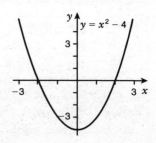

**Figure 24.4**

**Example 24.5.1**

Draw the graph of $y = x^2 - 2x - 1$ and use it to solve the equation $x^2 - 2x - 1 = 0$. Take values of $x$ from $-2$ to $4$.

| $x$ | $-2$ | $-1$ | 0 | 1 | 2 | 3 | 4 |
|-----|------|------|---|---|---|---|---|
| $y$ | 7 | 2 | $-1$ | $-2$ | $-1$ | 2 | 7 |

The diagram shows the graph of $y = x^2 - 2x - 1$. The $x$-coordinates of the points where the graph crosses the $x$-axis are $x = 2.4$ and $x = -0.4$. Thus, the solutions of the equation $x^2 - 2x - 1 = 0$ are, correct to one decimal place, $x = 2.4$ and $x = -0.4$.

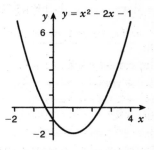

To use this method, you must have 0 on one side of the quadratic equation. The equation $x^2 = 3x + 5$, for example, would have to be rearranged as $x^2 - 3x - 5 = 0$ and then solved using the graph of $y = x^2 - 3x - 5$. Alternatively, you could draw the graph of $y = x^2$, a parabola, and the graph of $y = 3x + 5$, a straight line. The solutions of $x^2 - 3x - 5 = 0$ are then the $x$-coordinates of the points of intersection of the two graphs.

### EXERCISE 24.5

1 Solve each of these equations by drawing a graph, using the range of values of $x$ given in brackets. Where necessary, give your solutions correct to one decimal place.
   (a) $x^2 - 7 = 0$  ($-4$ to $4$)    (b) $x^2 - 2x = 0$  ($-2$ to $4$)
   (c) $x^2 - 4x + 3 = 0$  ($-2$ to $4$)   (d) $x^2 + x - 3 = 0$  ($-3$ to $3$)
   (e) $2x^2 - 3x - 7 = 0$  ($-4$ to $4$) (f) $x^2 = 3x + 2$  ($-2$ to $4$)

2 (a) Solve $x^2 - 4x + 4 = 0$, by drawing a graph and taking values of $x$ from $-2$ to $4$. In what way is this equation different from those in Question 1?
   (b) Draw the graph of $y = x^2 + x + 1$, taking values of $x$ from $-3$ to $3$. Explain why you cannot use your graph to solve the equation $x^2 + x + 1 = 0$.

# 24.6 Solution by factorizing

You can solve some quadratic equations by factorization. This depends on the fact that, if $a$ and $b$ are two numbers such that $ab = 0$, then either $a = 0$ or $b = 0$. So, if $(x - 2)(x - 4) = 0$, then $x - 2 = 0$ or $x - 4 = 0$, that is, $x = 2$ or $x = 4$. To use it, you must have 0 on one side of the equation and be able to factorize the expression on the other.

Thus, to solve the equation $x^2 - 4x + 3 = 0$, factorize to obtain $(x - 3)(x - 1) = 0$. Either $x - 3 = 0$ or $x - 1 = 0$, that is, $x = 3$ or $x = 1$.

---

**Example 24.6.1**

Solve these quadratic equations.
(a) $x^2 + 4x - 5 = 0$   (b) $x^2 - 2x = 0$   (c) $x^2 - 4x + 4 = 0$
(d) $x^2 - 4 = 0$       (e) $x^2 + 9x - 8 = 2x$

(a) $x^2 + 4x - 5 = 0$
$(x + 5)(x - 1) = 0$
Either $x + 5 = 0$ or $x - 1 = 0$, so $x = -5$ or $x = 1$.

(b) $x^2 - 2x = 0$
$x(x - 2) = 0$
Either $x = 0$ or $x - 2 = 0$, so $x = 0$ or $x = 2$.

(c) $x^2 - 4x + 4 = 0$
$(x - 2)(x - 2) = 0$
Either $x - 2 = 0$ or $x - 2 = 0$, so $x = 2$ or $x = 2$.

In this case there is only one root of the equation $x^2 - 4x + 4 = 0$, $x = 2$, although it is sometimes given as $x = 2$ repeated. See Exercise 24.5, Question 2(a). The parabola $y = x^2 - 4x + 4$ touches the $x$-axis.

(d) $x^2 - 4 = 0$
$(x - 2)(x + 2) = 0$
Either $x - 2 = 0$ or $x + 2 = 0$, so $x = 2$ or $x = -2$.

(e) Rearrange $x^2 + 9x - 8 = 2x$ to get 0 on the right-hand side. $x^2 + 9x - 8 = 2x$ so $x^2 + 7x - 8 = 0$
$(x - 1)(x + 8) = 0$
Either $x - 1 = 0$ or $x + 8 = 0$, so $x = 1$ or $x = -8$.

---

You can solve some problems by expressing the information as a quadratic equation and solving it.

## Example 24.6.2

The length of a rectangle is 4 cm greater than its width. The area of the rectangle is 96 cm$^2$. Find its width.

Let $x$ cm be the width of the rectangle. The length of the rectangle is, therefore, $(x + 4)$ cm and its area is $x(x + 4)$ cm$^2$. So $x(x + 4) = 96$.

$$x(x + 4) = 96$$
$$x^2 + 4x = 96$$
$$x^2 + 4x - 96 = 0$$
$$(x - 8)(x + 12) = 0.$$

Either $x - 8 = 0$ or $x + 12 = 0$,
  so $x = 8$ or $x = -12$.

As the width of a rectangle cannot be negative, $x = 8$. The width of the rectangle is 8 cm.

Check: The length of the rectangle is 12 cm and its area is 96 cm$^2$.

## EXERCISE 24.6

1 Write down the solutions to each of these quadratic equations.
  (a) $(x - 6)(x - 1) = 0$    (b) $(x + 4)(x - 3) = 0$
  (c) $(x + 7)(x - 7) = 0$    (d) $(x + 4)(x + 5) = 0$
  (e) $x(x - 6) = 0$          (f) $x(x + 8) = 0$

2 Solve these quadratic equations.
  (a) $x^2 - 3x + 2 = 0$      (b) $x^2 + x - 2 = 0$
  (c) $x^2 + 7x + 12 = 0$     (d) $x^2 - 25 = 0$
  (e) $x^2 - 10x + 25 = 0$    (f) $x^2 + 10x = 0$
  (g) $x^2 - x - 20 = 0$      (h) $x^2 - 64 = 0$
  (i) $x^2 - 7x = 0$          (j) $x^2 - x - 9 = 2x + 1$

3 The length of a rectangle is 3 cm greater than its width. The area of the rectangle is 28 cm$^2$. Find its length.

4 The square of a number is 27 more than 6 times the number. Find the number.

5 The sum of the square of a number and the number itself is 72. Find the number.

6 Two numbers have a difference of 2 and a product of 63. Find the numbers.

7 The product of two consecutive counting numbers is 56. Find the numbers.

## 24.7 Cubic equations

A cubic equation is one in which the highest power of $x$ is $x^3$. For example, $x^3 - 4x^2 - 2x - 3 = 0$, $x^3 = 3x + 7$ and $2x^3 + 5x^2 - 10 = 0$ are cubic equations.

You can solve simple cubic equations, such as $x^3 = 8$, in your head. A solution to this equation is $x = 2$, because $2^3 = 8$. Notice, however, that $x = -2$ is not a solution, as $(-2)^3 = -8$.

You can also find approximate solutions to equations like $x^3 = 10$, in which the right-hand side is not the cube of an integer. In this case, $x$ is the **cube root** of 10, written $\sqrt[3]{10}$. Using your calculator, you can find the value of $\sqrt[3]{10}$; it is 2.15, correct to two decimal places. Similarly, an approximate solution to $x^3 = -10$ is $x = -2.15$.

Sometimes you can use factorization to solve a cubic equation. For example, factorizing the left-hand side of the equation $x^3 - 4x = 0$, $x(x^2 - 4) = 0$. Factorizing again, $x(x - 2)(x + 2) = 0$ and so $x = 0$, $x = 2$ or $x = -2$.

Usually, though, you will have to solve cubic equations either by drawing an appropriate graph or by trial and improvement.

---

### Example 24.7.1

Draw the graph of $y = x^3 - 4x^2 - 2x + 3$ and use it to solve the equation $x^3 - 4x^2 - 2x + 3 = 0$. Take values of $x$ from $-2$ to 5.

| $x$ | $-2$ | $-1$ | 0 | 1 | 2 | 3 | 4 | 5 |
|---|---|---|---|---|---|---|---|---|
| $y$ | $-17$ | 0 | 3 | $-2$ | $-9$ | $-12$ | $-5$ | 18 |

The diagram shows the graph with equation $y = x^3 - 4x^2 - 2x + 3$. The $x$-coordinates of the points where the graph crosses the $x$-axis are $x = -1$, $x = 0.7$ and $x = 4.3$. Thus, the solutions of the equation $x^3 - 4x^2 - 2x + 3 = 0$ are $x = -1$, $x = 0.7$ and $x = 4.3$. The solution $x = -1$ is exact and the other two are correct to one decimal place.

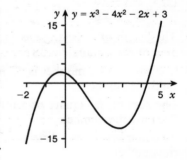

The above cubic equation has three solutions. This is the highest possible number of solutions a cubic equation can have, because a cubic graph can cross the $x$-axis at most three times. A cubic equation must have at least one solution, because a cubic graph must cross the $x$-axis at least once.

You can use **trial and improvement** to solve an equation to any degree of accuracy. For example, you can use it to find, correct to one decimal place, the higher positive solution to the equation $x^3 - 4x^2 - 2x + 3 = 0$, which was solved graphically in Example 24.7.1.

From the table of values and the graph, it is clear that there is a solution between $x = 4$ and $x = 5$. To determine which of these two values the solution is nearer to, evaluate $x^3 - 4x^2 - 2x + 3$ when $x = 4.5$.

| $x$ | 4 | 4.5 | 5 |
|---|---|---|---|
| $x^3 - 4x^2 - 2x + 3$ | $-5$ | 4.125 | 18 |

The change in the sign of $x^3 - 4x^2 - 2x + 3$ between $x = 4$ and $x = 4.5$ shows that the solution lies between these two values of $x$. The next step is to evaluate $x^3 - 4x^2 - 2x + 3$ when $x = 4.1$, 4.2 and so on until there is a sign change.

| $x$ | 4.1 | 4.2 | 4.3 | 4.4 |
|---|---|---|---|---|
| $x^3 - 4x^2 - 2x + 3$ | $-3.519$ | $-1.872$ | $-0.053$ | 1.944 |

The sign change occurs between $x = 4.3$ and $x = 4.4$, so the solution lies between these two values of $x$. The value of $x^3 - 4x^2 - 2x + 3$ when $x = 4.3$ is very close to 0, suggesting that, correct to one decimal place, the solution is $x = 4.3$ but you must evaluate $x^3 - 4x^2 - 2x + 3$ when $x = 4.35$ to confirm this.

| $x$ | 4.3 | 4.35 | 4.4 |
|---|---|---|---|
| $x^3 - 4x^2 - 2x + 3$ | $-0.053$ | 0.922 | 1.944 |

Thus, there is a solution between $x = 4.3$ and $x = 4.35$, so, correct to one decimal place, the solution is $x = 4.3$.

With trial and improvement, it is not necessary to have 0 on one side of the equation.

### Example 24.7.2

Solve the equation $x^3 + 2x = 57$, correct to one decimal place.

| $x$ | $x^3 + 2x$ | greater or smaller than 57 |
|-----|------------|----------------------------|
| 1 | 3 | smaller |
| 2 | 12 | smaller |
| 3 | 33 | smaller |
| 4 | 72 | greater |
| 3.5 | 49.875 | smaller |
| 3.6 | 53.856 | smaller |
| 3.7 | 58.053 | greater |
| 3.65 | 55.927… | smaller |

From the table, $x = 3.7$ (to one decimal place).

You could obtain a solution correct to two decimal places in the above example by evaluating $x^3 + 2x$ when $x = 3.66, 3.67$ etc. There is a solution between $x = 3.67$ and $x = 3.68$. Finally, evaluating $x^3 + 2x$ when $x = 3.675$ shows that, correct to two decimal places, $x = 3.68$.

### EXERCISE 24.7

1 Solve each of these equations by drawing a graph, using the range of values of $x$ given in brackets. Where necessary, give your solutions correct to one decimal place.
   (a) $x^3 - 7x^2 + 10x = 0$       (−1 to 6)
   (b) $x^3 - 2x^2 - 5x + 6 = 0$    (−3 to 5)
   (c) $x^3 - 7x - 5 = 0$           (−4 to 4)
   (d) $x^3 - x^2 - 4x + 4 = 0$     (−3 to 3)
   (e) $x^3 - 4x^2 + 4x = 0$       (−2 to 4)
   (f) $x^3 - 2x^2 - 2 = 0$         (−2 to 3)

2 There is a solution to the equation $x^3 + x - 200 = 0$ between $x = 5$ and $x = 6$. Find this solution correct to one decimal place.

3 There is a solution to the equation $x^3 + 2x^2 - 120 = 0$ between $x = 4$ and $x = 5$. Find this solution correct to one decimal place.

4 Find, correct to one decimal place, a positive solution to the equation $x^3 - 3x - 15 = 0$.

5 Solve the equation $x^3 + 6x = 15$ correct to one decimal place.

6 Solve the equation $x^3 + 4x^2 = 32$ correct to one decimal place.

**answers**

### Exercise 1.1 (page 2)

1  (a) fifty (b) five hundred (c) five thousand (d) five

2  (a) 17 (b) 70 (c) 97 (d) 546 (e) 603 (f) 810 (g) 450 (h) 10 000 (i) 8934 (j) 6480 (k) 3006

3  (a) fifty two (b) eight hundred and seventy one (c) five thousand, six hundred and twenty four (d) nine hundred and eighty (e) seven thousand and one (f) thirty five thousand and thirteen (g) two hundred and forty one thousand and one (h) one million, one thousand, three hundred and twelve.

4  (a) 975, 579 (b) 9721, 1279

5  (a) three hundred and forty two thousand, seven hundred and eighty five (b) three million, seven hundred and eighty three thousand, one hundred and ninety four (c) seventeen million, twenty one thousand, two hundred and nine (d) three hundred and five million, two hundred and thirteen thousand and ninety seven

6  (a) 516 219 (b) 206 024 (c) 21 437 869 (d) 7 604 013

### Exercise 1.2 (page 4)

1  (a) 29 (b) 48 (c) 54 (d) 61

2  (a) 94 (b) 114

3  (a) 64 (b) 14 (c) 516 (d) 85 (e) 66 (f) 78

4  (a) 18 (b) 135

5  (a) 45 (b) 301 (c) 2574 (d) 3556 (e) 297 (f) 336 (g) 182 (h) 441

6  2484

7  (a) 75 (b) 282 (c) 243 (d) 219

8  (a) 2 r 2 (b) 4 r 3 (c) 12 r 5

### Exercise 1.3 (page 6)

1  (a) 35 (b) 23 (c) 8 (d) 1 (e) 1 (f) 4 (g) 6 (h) 23 (i) 26 (j) 8 (k) 3 (l) 70 (m) 14 (n) 39 (o) 3 (p) 22 (q) 32 (r) 17

## Exercise 1.4 (page 6)

1 993

2 168

3 1504

4 23

5 612

6 19, 3 cm

7 32

8 189 miles

9 10 220

10 23, 4

## Exercise 1.5 (page 9)

1 (a) 34, 112, 568, 4366 (b) 67, 741

2 1, 4, 9, 16, 25, 36, 49, 64, 81, 100

3 1, 3, 6, 10, 15, 21, 28, 36, 45, 55

4 4, 6, 8, 9, 10, 12, 14, 15, 16, 18

5 1, 8, 27, 64, 125, 216, 343, 512, 729, 1000

6 (a) 25, 36 (b) 36 (c) 20, 25, 28, 33, 36

7 odd

8 (a) e.g. $3 + 6 = 9, 6 + 10 = 16, 10 + 15 = 25$ (b) They are squares

9 9, 36, 100, 225; they are square numbers

10 (a) e.g. 16 (b) e.g. 24 (c) e.g. 36

## Exercise 1.6 (page 10)

1 7, 14, 21, 28, 35

2 (a) 5, 10, 15, 20, 25, 30, 35, 40, 45, 50 (b) ends in 5

3 1, 2, 4, 5, 10, 20

4 (a) $1 + 2 + 4 + 7 + 14 = 28$ (b) same as original number

5 1, 23

6 (a) e.g. 9, 16, 25, 36, 49 (b) square numbers

7 23, 29, 31, 37

8 Other even numbers have factor 2

9 10

10 24

11 2, 3, 5

12 5, 7

13 e.g. 12

## Exercise 2.1 (page 13)

1  10

2  6°

3  $17\frac{1}{2}°$

4  330°

5  (a) 90° (b) 270°

6  (a) 45° (b) 315°

7  (a) obtuse (b) reflex (c) acute (d) obtuse (e) reflex

8  (a) obtuse (b) reflex (c) reflex (d) acute

9  ∠L or ∠MLN; ∠N or ∠LNM; ∠A or ∠BAD; ∠CBD.

## Exercise 2.2 (page 15)

1  (a) 50° (b) 132° (c) 315° (d) 111° (e) 231°

## Exercise 2.3 (page 17)

1  (a) 67° (b) 23° (c) 111° (d) 58° (e) 152° (f) 72° (g) 63°
   (h) 202° (i) 129°, 51°

2  No 37° + 97° + 45° = 179° not 180°.

## Exercise 2.4 (page 19)

1  (a) (i) $d$ (ii) $b$ (b) (i) $f$ (ii) $h$ (c) (i) $k$ (ii) $i$

2  (a) $a=57°$, $b=57°$, $c=123°$ (b) $d=98°$, $e=82°$ (c) $f=78°$, $g=28°$,
   $h=74°$ (d) $i=66°$, $j=114°$, $k=115°$, $l=115°$ (e) $m=59°$, $n=41°$, $p=260°$
   (f) $q=65°$, $r=65°$

## Exercise 2.5 (page 21)

1  063°, 158°, 248°, 305°

3  (a) 270° (b) 045° (c) 225° (d) 315°

4  (a) 149° (b) 293° (c) 204°

5  319°

6  025°

## Exercise 3.1 (page 24)

1  (a) $\frac{4}{5}, \frac{7}{10}, \frac{5}{12}$ (b) $\frac{1}{3}, \frac{3}{10}, \frac{7}{12}$

2  $\frac{1}{6}$

3  $\frac{4}{7}$

4  $\frac{5}{6}, \frac{3}{8}$

5  (a) $\frac{2}{6}, \frac{3}{9}, \frac{4}{12}, \frac{5}{15}, \frac{6}{18}$ (b) $\frac{6}{8}, \frac{9}{12}, \frac{12}{16}, \frac{15}{20}, \frac{18}{24}$ (c) $\frac{4}{10}, \frac{6}{15}, \frac{8}{20}, \frac{10}{25}, \frac{12}{30}$ (d) $\frac{10}{18}, \frac{15}{27}, \frac{20}{36}, \frac{25}{45}, \frac{30}{54}$

6  $\frac{2}{3} = \frac{16}{24}$ and $\frac{5}{8} = \frac{15}{24}$ so $\frac{2}{3} > \frac{5}{8}$

## Exercise 3.2 (page 25)

1 (a) $\frac{4}{5}$ (b) $\frac{2}{3}$ (c) $\frac{7}{9}$ (d) $\frac{5}{16}$ (e) $\frac{5}{6}$ (f) $\frac{5}{8}$ (g) $\frac{2}{9}$ (h) $\frac{2}{3}$

2 (a) $\frac{1}{2}, \frac{7}{12}, \frac{5}{8}$ (b) $\frac{7}{10}, \frac{3}{4}, \frac{4}{5}$

3 (a) $\frac{4}{8} = \frac{2}{4}$ (b) $\frac{6}{18} = \frac{2}{6}$ (c) $\frac{20}{24} = \frac{5}{6}$

4 (a) $\frac{30}{40} = \frac{6}{8}$ (b) $\frac{24}{36} = \frac{4}{6}$ (c) $\frac{18}{27} = \frac{2}{3}$

5 (a) $\frac{2}{3}$ (b) $\frac{3}{4}$ (c) $\frac{1}{4}$ (d) $\frac{4}{5}$ (e) $\frac{6}{7}$ (f) $\frac{8}{15}$ (g) $\frac{1}{4}$ (h) $\frac{3}{5}$

6 (a) $\frac{1}{3}$ (b) $\frac{1}{2}$ (c) $\frac{1}{3}$ (d) $\frac{3}{5}$ (e) $\frac{5}{6}$ (f) $\frac{3}{5}$ (g) $\frac{2}{3}$ (h) $\frac{3}{5}$

## Exercise 3.3 (page 27)

1 (a) 5 (b) 7 (c) 6 (d) 7

2 (a) $\frac{3}{1}$ (b) $\frac{12}{1}$

3 (a) $2\frac{3}{5}$ (b) $4\frac{3}{4}$ (c) $4\frac{1}{6}$ (d) $1\frac{7}{8}$ (e) $6\frac{2}{3}$ (f) $3\frac{3}{5}$ (g) $2\frac{1}{8}$ (h) $3\frac{2}{9}$

4 (a) $3\frac{1}{2}$ (b) $3\frac{1}{2}$ (c) $1\frac{2}{3}$ (d) $2\frac{1}{2}$

5 (a) $\frac{14}{5}$ (b) $\frac{14}{3}$ (c) $\frac{7}{2}$ (d) $\frac{23}{4}$

## Exercise 3.4 (page 29)

1 (a) $\frac{4}{5}$ (b) $\frac{6}{7}$ (c) $\frac{2}{3}$ (d) $\frac{2}{3}$ (e) $1\frac{2}{9}$ (f) $1\frac{1}{4}$ (g) 1 (h) $1\frac{3}{5}$

2 (a) $\frac{3}{5}$ (b) $\frac{2}{7}$ (c) $\frac{1}{4}$ (d) $\frac{2}{3}$

3 (a) $\frac{11}{15}$ (b) $\frac{5}{8}$ (c) $\frac{2}{3}$ (d) $1\frac{1}{2}$ (e) $1\frac{4}{15}$ (f) $\frac{11}{18}$ (g) $1\frac{7}{24}$ (h) $1\frac{1}{3}$

4 (a) $6\frac{1}{20}$ (b) $5\frac{9}{10}$ (c) $7\frac{4}{5}$ (d) $5\frac{7}{12}$

5 (a) $2\frac{5}{8}$ (b) $2\frac{1}{6}$ (c) $1\frac{1}{12}$ (d) $2\frac{5}{6}$

## Exercise 3.5 (page 31)

1 (a) $3\frac{1}{3}$ (b) 12 (c) $7\frac{1}{2}$ (d) 35

2 (a) $\frac{15}{28}$ (b) $\frac{7}{12}$ (c) $\frac{3}{4}$ (d) $\frac{2}{15}$

3 (a) $4\frac{3}{8}$ (b) $5\frac{1}{4}$ (c) 8 (d) $1\frac{11}{24}$ (e) 2 (f) $1\frac{1}{4}$ (g) $\frac{3}{5}$ (h) 14

4 (a) 5 (b) 35 (c) 27 (d) 33

5 (a) 32 kg (b) 21 cm (c) 48 litres

## Exercise 3.6 (page 32)

1 (a) 6 (b) $5\frac{1}{3}$ (c) 9 (d) $3\frac{1}{3}$

2 (a) $2\frac{2}{15}$ (b) $\frac{5}{9}$ (c) $1\frac{1}{6}$ (d) $\frac{27}{40}$ (e) $\frac{2}{15}$ (f) $\frac{7}{8}$ (g) $2\frac{4}{9}$ (h) $\frac{16}{21}$

3 (a) $1\frac{1}{3}$ (b) $\frac{33}{40}$ (c) 6 (d) $\frac{2}{3}$ (e) $2\frac{1}{40}$ (f) $\frac{7}{9}$ (g) $\frac{5}{24}$ (h) $3\frac{9}{10}$

## Exercise 3.7 (page 33)

1 (a) $\frac{2}{15}$ (b) $\frac{5}{8}$ (c) $\frac{5}{14}$ (d) $\frac{2}{5}$ (e) $\frac{5}{12}$ (f) $\frac{2}{7}$ (g) $\frac{3}{5}$ (h) $\frac{3}{16}$ (i) $\frac{7}{20}$ (j) $\frac{1}{12}$

2 $\frac{2}{3}$

3 $\frac{7}{10}$

4 $\frac{4}{15}$

5 $\frac{5}{16}$

6 $3\frac{2}{5}$

7 $\frac{7}{12}$

8 240°

9 9

## Exercise 4.1 (page 37)

1 81°

2 73°, 34°

3 63°

4 60°, 120°

5 31°

6 45°

7 58°, 61°

8 66°

9 68°, 68°

10 $x = 74°, y = 32°$ or $x = 53°, y = 53°$

11 140°, 20°, 20°

## Exercise 4.2 (page 39)

8 Cannot be done – arcs do not cross. The sum of the lengths of any two sides must be greater than the length of the third side.

## Exercise 4.3 (page 41)

1 112°

2 two from rectangle, parallelogram, rhombus

3 square, rhombus, kite

4 0, 1, 2, 4

5 1, 2, 4

## Exercise 4.4 (page 44)

1 126°

2 155°

3 127°

4 1800°

5 2340°

6 108°

7 144°

8  160°

9  71°

10  60°

11  (a) 30° (b) 150°

12  9

13  30

## Exercise 4.5 (page 46)
A and E, B and H, C and G, D and J.

## Exercise 4.6 (page 47)
2  Each interior angle of a rectangle is 90°, so four corners fit to make 360°.

3  Each interior angle of a regular nonagon, 140°; 140 does not divide into 360 exactly.

5  Each interior angle of an equilateral triangle is 60° and of a regular 12-gon is 150°. $60 + 2 \times 150 = 360$, so the two shapes might tessellate. They do.

## Exercise 5.1 (page 50)
1  (a) $\frac{3}{1000}$ (b) 30 (c) $\frac{3}{10}$ (d) $\frac{3}{100}$

2  (a) 3 (b) $\frac{8}{10}$ (c) $\frac{9}{100}$ (d) $\frac{1}{100}$

3  (a) 2 (b) 7

4  (a) 3 (b) 0

5  (a) 0.901 (b) 0.72 (c) 0.01 (d) 0.909

6  5.632

7  0.678

8  1.203, 1.23, 1.302, 1.32, 2.31

9  0.019, 0.109, 0.190, 0.901, 0.910

10  (a) 4.76 < 6.47 (b) 5.72 > 5.27 (c) 0.39 < 0.4 (d) 0.03 > 0.029

## Exercise 5.2 (page 52)
1  (a) $\frac{3}{5}$ (b) $\frac{1}{10}$ (c) $\frac{12}{25}$ (d) $\frac{39}{100}$ (e) $\frac{9}{100}$ (f) $\frac{179}{1000}$ (g) $\frac{1}{8}$ (h) $\frac{3}{1000}$

2  (a) $3\frac{4}{5}$ (b) $4\frac{11}{50}$ (c) $54\frac{3}{4}$ (d) $23\frac{6}{125}$

3  (a) 0.3 (b) 0.71 (c) 0.07 (d) 0.387 (e) 0.017 (f) 0.009 (g) 0.4 (h) 0.95 (i) 0.68 (j) 0.22 (k) 0.474 (l) 0.035

4  (a) 6.1 (b) 9.03 (c) 5.041 (d) 1.45 (e) 8.32 (f) 12.185

## Exercise 5.3 (page 53)
1  (a) 7.9 (b) 13.12 (c) 10.3 (d) 24.07 (e) 6.3 (f) 16.8 (g) 7.17 (h) 14.83 (i) 10.7 (j) 2.2 (k) 11.88 (l) 3.37 (m) 3.851 (n) 3.488 (o) 12.085

2  0.77

3  34.8

4  10.28 s

## Exercise 5.4 (page 56)

1  (a) 28.2 (b) 89 (c) 146.96 (d) 1.209

2  (a) 82.1 (b) 6 (c) 4.7 (d) 7.02 (e) 283 (f) 0.041 (g) 521.8
   (h) 83 (i) 40 (j) 0.07 (k) 46 360 (l) 3900

3  (a) $0.21 \times 100 = 21$ (b) $6.71 \times 10 = 67.1$ (c) $7.98 \times 1000$
   $= 7980$

4  (a) 0.63 (b) 0.3 (c) 0.08 (d) 0.03 (e) 0.021 (f) 0.0006

5  (a) 5.18 (b) 0.1148 (c) 2.516 (d) 0.03948 (e) 344.1 (f) 1.127

6  10.56 pints

7  36 litres

8  832 cm

## Exercise 5.5 (page 58)

1  (a) 2.3 (b) 6.41 (c) 0.397 (d) 0.067

2  (a) 3.75 (b) 45.72 (c) 0.0093 (d) 0.713 (e) 0.034 (f) 6.528
   (g) 0.025 (h) 0.0037 (i) 0.00003 (j) 0.172 (k) 0.0338
   (l) 0.000071

3  (a) $3.7 \div 10 = 0.37$ (b) $0.9 \div 100 = 0.009$ (c) $7.9 \div 100 = 0.079$

4  (a) 19.96 (b) 28.5 (c) 0.156 (d) 0.91 (e) 3.68 (f) 0.036 (g) 0.875 (h) 0.006545

5  (a) 0.9 (b) 15 (c) 80 (d) 0.03 (e) 300 (f) 50 (g) 24.6 (h) 37.6

6  4.62 m

7  0.021 cm

8  25

## Exercise 5.6 (page 60)

1  (a)  0.25  terminating  (b)  0.$\dot{3}$  recurring  (c)  0.375  terminating
   (d) 0.$\dot{4}$ recurring (e) 0.875 terminating

2  (a) 0.8$\dot{1}$ (b) 0.0$\dot{9}$ (c) 0.41$\dot{6}$ (d) 0.08$\dot{3}$ (e) 0.5$\dot{7}$ (f) 0.08$\dot{1}$ (g) 0.34$\dot{8}$ (h) 0.52$\dot{7}\dot{0}$

3  (a) 0.$\dot{8}$ (b) 0.7$\dot{1}\dot{2}$ (c) 0.3$\dot{6}\dot{7}$ (d) 0.45$\dot{6}8\dot{9}$ (e) 18.0$\dot{1}\dot{8}$ (f) 0.$\dot{5}64\dot{3}$ (g) 9.3$\dot{9}$ (h) 263.$\dot{6}\dot{3}$

4  (a) $\frac{7}{11} > \frac{8}{13}$ (b) $\frac{14}{19} < \frac{17}{23}$ (c) $\frac{6}{13} < \frac{7}{15}$ (d) $\frac{13}{17} > \frac{16}{21}$

5  (a)  5.77,   $5\frac{17}{22}$,   $5\frac{7}{9}$,   5.78,   5.8   (b)  3.01,   3.09,   $3\frac{1}{11}$,   $3\frac{2}{21}$,   3.1
   (c) 4.083, $4\frac{1}{12}$, 4.084, $4\frac{2}{23}$, 4.085

6  0.516, $\frac{129}{250}$

7  (a) 0.$\dot{1}4285\dot{7}$, 0.$\dot{2}8571\dot{4}$, 0.$\dot{4}2857\dot{1}$, 0.$\dot{5}7142\dot{8}$, 0.$\dot{7}1428\dot{5}$, 0.$\dot{8}5714\dot{2}$ (b) 6
   (c) Same six digits in each recurring group.

**8** (a) 2, 4, 5, 8, 10, 16, 20, 25 (b) Each denominator is either a power of 2, a power of 5 or a product of powers of 2 and 5.

## Exercise 6.1 (page 65)
**1** 14, 4, 5, 9

**2** 8, 7, 5, 4, 2

**6** 20, 21, 60, 65

**10** 13, 33, 59, 10, 7, 63

## Exercise 6.2 (page 69)
**1** 120°, 105°, 90°, 15°, 30°

**2** 66°, 54°, 114°, 84°, 42°

**3** 189°, 104°, 66°

**4** (a) 25 min (b) 105 min (c) $\frac{11}{72}$

## Exercise 6.3 (page 71)
**1** (b) 6.5 cm

**2** (b)(i) 17°C (ii) 15–16 min

**3** (b) 97 cm

**4** (b) 120 million, 1970

## Exercise 6.4 (page 75)
**1** (c) low positive

**2** (c) high negative (d) 27

**3** no correlation

## Exercise 6.5 (page 76)
**1** discrete: (a), (d), (e), (h), (j) continuous: (b), (c), (f), (g), (i)

## Exercise 6.6 (page 79)
**1** (a) 14, 7, 5, 3, 1

**2** (a) 10, 7, 5, 3

**4** (a) 4, 6, 5, 3, 2

## Exercise 7.1 (page 81)
**1** −7°C

**2** −9

**3** (a) −3 < −1 (b) 0 > −2 (c) −12 > −15

**4** (a) 2°C (b) −8°C

**5** −4, −3 −1, 0, 2, 5

## Exercise 7.2 (page 83)

1   5°C

2   −4°C

3   11

4   7

5   (a) 1 (b) −3 (c) 0 (d) −3 (e) −5 (f) −1 (g) −9 (h) −5 (i) 0
    (j) −4

6   (a) 1 (b) 7 (c) −9 (d) −1 (e) 6 (f) 0 (g) −10 (h) −7

## Exercise 7.3 (page 86)

1   (a) −8 (b) 12 (c) 0 (d) −12 (e) 16

2   (a) −20 (b) 24 (c) −21 (d) 0 (e) 36 (f) −64 (g) 35 (h) 0
    (i) −72

3   (a) −6 (b) 8 (c) −3 (d) 5 (e) 0 (f) −6 (g) −4 (h) 4 (i) −7

## Exercise 7.4 (page 86)

1   (a) −408 (b) 56 (c) −139 (d) −26 (e) −5.1 (f) −1.6 (g) −2.66 (h) 4.8

## Exercise 8.1 (page 89)

1   $A(4,2)$ $B(1,3)$ $C(2,2)$ $D(0,1)$ $E(3,0)$

3   (c) $(6,1)$

4   $A(−4,2)$ $B(3,1)$ $C(−3,−1)$ $D(2,−3)$ $E(0,−3)$

6   (b) kite

7   (b) $(1,−2)$

8   $(−1,3)$

## Exercise 8.2 (page 92)

1   (a)(i) 0.4 (ii) 1.4 (b)(i) 12 (ii) 18

2   (a)(i) 45 km (ii) 67 km (b)(i) 22 miles (ii) 55 miles

3   (a)(i) €55 (ii) €38 (b)(i) £52 (ii) £33

4   (a) 1315 (b) 45 mins (c) 40 mph (d) 48 mph

5   (b) 81 gallons (c) 27 gallons

6   (b) $33 (c) £27

7   (b) 18 km/h (c) 20 km/h

8   (a) 40 60 80 100 120 140 (c) 104 mins

## Exercise 8.3 (page 96)

1   $P$ $x=2$, $Q$ $x=−3$, $R$ $y=3$, $S$ $y=−2$

3   (a) $x < −2$ (b) $−1 < y \leq 2$

## Exercise 9.1 (page 99)

1  (a) 7.9 cm (b) 8.5 m (c) 3.24 kg (d) 4.125 l (e) 1800 m
   (f) 0.4 km (g) 94 mm (h) 940 ml (i) 2300 g (j) 3720 kg

2  (a) 5.4 cm (b) 54 mm

3  120

4  55.66 m

5  5.37 kg

6  33 days

## Exercise 9.2 (page 100)

1  (a) 96 in (b) 5 gallons (c) 3 lb (d) 56 pints (e) 8 yd (f) 5280 yd (g) 70 in (h) 55 oz
   (i) 3 ft 7 in (j) 3 lb 9 oz

2  6 ft 2 in

3  8 in

4  14 gallons

5  4920 pt

6  1980

## Exercise 9.3 (page 101)

1  (a) 15 cm (b) 14 in (c) 12 ft (d) 72 km (e) 6 oz (f) 11 lb (g) 21 pt (h) 12 gallons
   (i) 60 miles (j) 8 l

2  80 cm

3  17 oz

4  42 in

5  8 gallons

6  187 lb

7  192 km

8  5 ft 10 in

9  155 miles

10  5

11  16

## Exercise 9.4 (page 102)

1  (a)(i) metres (ii) feet (b)(i) grams (ii) ounces (c)(i) litres (ii) pints (d)(i) kilometres
   (ii) miles (e)(i) litres (ii) gallons (f)(i) kilograms (ii) pounds

2  (a) grams (b) litres (c) grams (d) litres (e) millilitres

## Exercise 10.1 (page 104)

1  17 cm

2  27 cm

3  37.2 cm

4  23 cm

5  36 cm

6  (a) 328 m (b) 366 m

7  (a) 14 (b) 16

8  20

9  (a) 10 (b) 9 (c) 9

10  (a) 11 (b) 13

11  11

12  (a) 14 cm (b) 10 cm, 12 cm, 14 cm (c) 14 cm

## Exercise 10.2 (page 106)

1  (a) 6 cm$^2$ (b) 4 cm$^2$ (c) $3\frac{1}{2}$ cm$^2$ (d) 4 cm$^2$ (e) 6 cm$^2$ (f) 5 cm$^2$

2  (a) 6 (b) 6 (c) 9 (d) 9 (e) 16 (f) 7

3  28 cm$^2$

## Exercise 10.3 (page 108)

1  84 cm$^2$

2  48 m$^2$

3  81 cm$^2$

4  (a) 1.38 m$^2$ (b) 13 800 cm$^2$

5  (a) 1 m$^2$ (b) 10 000 cm$^2$

6  (a) 22 cm$^2$ (b) 50 cm$^2$ (c) 31 cm$^2$

7  (a) 44 cm$^2$ (b) 38 cm$^2$

8  (a) 54 m$^2$ (b) 20 m$^2$ (c) 7 litres

9  9 m$^2$

10  36 m$^2$

11  10 000 m$^2$

12  42

## Exercise 10.4 (page 112)

1  (a) 24 cm$^2$ (b) 6 cm$^2$ (c) 16 cm$^2$ (d) 36 cm$^2$ (e) 6 cm$^2$ (f) 35 cm$^2$ (g) 47 cm$^2$ (h) 28 cm$^2$

2  (a) 5.4 cm$^2$ (b) 3.8 cm$^2$

3  (a) 33 cm$^2$ (b) 50 cm$^2$

4  $5\frac{1}{2}$ cm$^2$

5  40 cm$^2$

6  60 cm$^2$

## Exercise 10.5 (page 115)

**1** (a) $55 \, \text{cm}^2$ (b) $66.5 \, \text{cm}^2$ (c) $24 \, \text{cm}^2$ (d) $12 \, \text{cm}^2$

**2** $6 \, \text{cm}^2$

**3** (a) 10 (b) 10.5 (c) 16

**4** (a) $37 \, \text{cm}^2$ (b) $53.5 \, \text{cm}^2$

**5** 6

**6** 8

## Exercise 11.1 (page 118)

**1** (a) $x + 2$ (b) $x - 1$ (c) $4x$ (d) $\frac{1}{2}x$

**2** (a) $p + q$ (b) $pq$

**3** $\dfrac{8}{x}$

**4** $a - b$

**5** $30 - b$

**6** (a) $6x$ (b) $\frac{1}{3}y$ (c) $xy$ (d) $3xy$ (e) $\dfrac{x}{y}$ (f) $\dfrac{5x}{y}$ (g) $\frac{1}{3}xy$ (h) $xyz$

**7** $7w$

**8** $\dfrac{1}{100}d$ or $\dfrac{d}{100}$

**9** $np$

**10** $\dfrac{a}{b}$

## Exercise 11.2 (page 120)

**1** $5l$

**2** (a) $2a + 2b$ (b) $ab$

**3** $2x + 4y$

**4** $4a + b$

**5** (a) $2x$ (b) $4x$ (c) $7x$ (d) $3x$ (e) $9x$ (f) $y$ (g) $-2xy$ (h) $-abc$ (i) $9x + 5y$ (j) $2x + 3$ (k) $7x + 3$ (l) $5x$ (m) $9xy + x - 3$ (n) $7a - 2b + 6$ (o) $8a + ab + 6$ (p) $6a$ (q) $2 - 4pq + r$ (r) $p + q$

## Exercise 11.3 (page 121)

**1** (a) 19 (b) 5 (c) 1 (d) $-5$

**2** (a) 15 (b) 6 (c) 51 (d) $-16$

**3** (a) 16 (b) 60 (c) 300 (d) 34

## Exercise 11.4 (page 122)

**1** (a) 9 (b) 10 (c) 45 (d) $-3$

**2** (a) 16 (b) 13 (c) 37 (d) 40

**3** (a) 85 (b) 22 (c) 76 (d) 9

### Exercise 11.5 (page 124)

1 (a) $6x + 18$ (b) $4x - 4$ (c) $10x + 15$ (d) $6x - 10$ (e) $-4x - 12$
(f) $-7x + 14$ (g) $-40x - 32$ (h) $-18x + 12$ (i) $x^2 + 2x$
(j) $4x^2 - 3x$ (k) $6x^2 - 3x$ (l) $-15x^2 - 20x$

2 (a) $9x - 21$ (b) $10 - 10x$ (c) $9x - 7$ (d) $x + 22$ (e) $18x - 13$
(f) $21 - x$ (g) $26x$ (h) 6 (i) $x^2 - 4x - 24$ (j) $x^2 + 24$
(k) $x^2 + x - 12$ (l) $6x^2 + 6x + 28$ (m) $23x + 2$ (n) $18x + 1$
(o) $-6$ (p) $1 + 5x$ (q) $-5$ (r) $x$

### Exercise 11.6 (page 125)

1 (a) $3(x + 3)$ (b) $7(2x - 3)$ (c) $5(4x + 3)$ (d) $3(2x - 5y)$
(e) $a(x - 7)$ (f) $x(x + 4)$ (g) $6(2x - 3)$ (h) $3(5x - 4)$
(i) $2x(2a + 3b)$ (j) $3x(3x + 4)$ (k) $4x(5 - 2x)$ (l) $2bx(2 - x)$
(m) $xy(x + y)$ (n) $3ax(2x - 5a)$ (o) $4xy(2 + 3y)$

### Exercise 11.7 (page 126)

1 (a) $m^3$ (b) $n^5$ (c) $5p^5$ (d) $3q^6$

2 (a) 243 (b) 135

3 (a) 1 000 000 (b) 70 000

4 (a) $-64$ (b) $-3072$

5 (a) $-500$ (b) 3750

### Exercise 11.8 (page 128)

1 (a) $x^7$ (b) $x^5$ (c) $x^{12}$ (d) $x^6$ (e) $x^5$ (f) $x^{10}$ (g) $x^{10}$ (h) $x^3 y^4$ (i) $x^4$
(j) $x^5$ (k) $\dfrac{x^5}{y^3}$ (l) $x^4$ (m) $x$ (n) $x^{10}$ (o) $x^2$ (p) $x^9$

### Exercise 11.9 (page 130)

1 (a) $6x^5$ (b) $20x^3$ (c) $14x^5 y^6$ (d) $9x^8$ (e) $6x^9$ (f) $8x^7 y^5$ (g) $3x^2$ (h) $7x^3$ (i) $5x^2 y^2$

2 (a) $3xy^3$ (b) $6x^5 y^2$ (c) $\dfrac{5x^4 y^3}{z^2}$ (d) $\dfrac{1}{a^5}$ (e) $\dfrac{8}{x^4}$ (f) $\dfrac{4x^4}{y^2}$ (g) $\dfrac{y^4}{3x}$ (h) $5y^4$ (i) $36x^2$
(j) $25x^2 y^2$ (k) $x^{10} y^6$ (l) $25x^6 y^8$ (m) $1000x^3$
(n) $64x^{15}$ (o) $125x^{12} y^9$

### Exercise 12.1 (page 134)

1 (a) 30 (b) 50 (c) 80 (d) 750 (e) 8470 (f) 7400

2 (a) 700 (b) 800 (c) 400 (d) 4400 (e) 3000 (f) 19 500

3 (a) 8000 (b) 3000 (c) 6000 (d) 38 000 (e) 42 000 (f) 100 000

4 61 300

5 (a) 242 000 (b) 200 000

6 (a) 7550 (b) 7649

## Exercise 12.2 (page 136)

1 (a) 9 (b) 21 (c) 68 (d) 260 (e) 400

2 (a) 9.9 (b) 0.9 (c) 18.0 (d) 76.9 (e) 400.0

3 (a) 4.25 (b) 8.47 (c) 0.75 (d) 0.06

4 (a) 8.678 (b) 7.609 (c) 13.420 (d) 0.009

5 10

6 2.1

7 17.41

8 0.429

9 7.25

10 5.845

## Exercise 12.3 (page 138)

1 (a) 20 (b) 400 (c) 0.2 (d) 0.007 (e) 0.09

2 (a) 8700 (b) 740 (c) 6.9 (d) 0.085 (e) 0.0040

3 (a) 23 700 (b) 376 (c) 13.8 (d) 8.93 (e) 0.0844 (f) 0.000 437 (g) 0.0902 (h) 0.003 70

4 2000

5 5.9

6 (a) 11.3 (b) 0.480

7 0.0909

8 (a) 600 (b) 600 (c) 600

9 (a) 0.06 (b) 0.060 (c) 0.0600

## Exercise 12.4 (page 139)

1 (a) 8, 8.9386 (b) 90, 98.7953 (c) 5, 5.32 (d) 10 000, 11 811.38 (e) 5, 5.57 (f) 4, 3.29 (g) 8400, 8194.4896 (h) 12, 13.6 (i) 21, 23.3531 (j) 0.3, 0.341 (k) 200, 197 (l) 16, 13.3104 (m) 0.000 14, 0.000159 (n) 20, 21.8 (o) 0.2, 0.248 (p) 40, 50.8 (q) 0.02, 0.0228 (r) 2, 1.52 (s) 4, 3.86 (t) 100, 122

## Exercise 12.5 (page 140)

1 41, 4

2 8, 4 cm

3 4

4 3

5 6

6 6

7 10

8 8

## Exercise 12.6 (page 142)

1  324.5 km, 325.5 km

2  12.65 cm, 12.75 cm

3  9.845 s, 9.855 s

4  (a) 0.1 kg (b) 10.55 kg, 10.65 kg (c) 0.05 kg

5  (a) 0.01 s (b) 0.005 s

## Exercise 13.1 (page 144)

1  (a) 5 (b) 11 (c) 9 (d) 12 (e) 7 (f) 4 (g) 2 (h) 4 (i) 7

2  (a) 13 (b) 17 (c) 7 (d) 27 (e) 6

3  (a) 4 (b) 6 (c) 4 (d) 12

## Exercise 13.2 (page 146)

1  (a) 3 (b) 8 (c) 6 (d) 10 (e) $-5$ (f) $\frac{3}{5}$ (g) 0 (h) $2\frac{1}{3}$ (i) 0 (j) $-5$
(k) 2 (l) $-6$

## Exercise 13.3 (page 148)

1  (a) 4 (b) 3 (c) $\frac{1}{2}$ (d) $-1$ (e) $2\frac{1}{3}$ (f) $-\frac{3}{4}$ (g) 4 (h) 0 (i) $2\frac{1}{2}$ (j) $\frac{5}{6}$
(k) $-2$ (l) $-2\frac{2}{3}$

2  (a) 4 (b) 5 (c) $\frac{1}{2}$ (d) $-2$ (e) $1\frac{2}{3}$ (f) $-\frac{2}{5}$ (g) $-2\frac{1}{2}$ (h) $\frac{2}{3}$ (i) 0 (j) $-3$ (k) $-\frac{2}{3}$ (l) $2\frac{3}{4}$

3  (a) 5 (b) $-5$ (c) 2 (d) $\frac{3}{5}$ (e) $-2$ (f) $2\frac{1}{3}$ (g) $-\frac{1}{2}$ (h) $-1\frac{1}{2}$ (i) 1 (j) $\frac{4}{5}$ (k) $-1$ (l) 0 (m) $2\frac{3}{5}$ (n) $-\frac{2}{5}$ (o) 3 (p) $-4$

## Exercise 13.4 (page 149)

1  (a) 3 (b) 2 (c) $-3$ (d) $\frac{2}{3}$ (e) 12 (f) $1\frac{1}{2}$ (g) $1\frac{1}{3}$ (h) $-\frac{3}{4}$ (i) $-2$ (j) 2 (k) $-2\frac{2}{3}$ (l) $\frac{7}{8}$
(m) $1\frac{1}{4}$ (n) $-\frac{1}{4}$ (o) $1\frac{1}{2}$ (p) 5 (q) $-7$ (r) $\frac{1}{2}$ (s) 0 (t) $-2$

## Exercise 13.5 (page 151)

1  12

2  8

3  19

4  11

5  35

6  32, 8

7  10 cm

8  33°

9  8, 9, 10

10  5

## Exercise 13.6 (page 153)

1 (a) $x < 3$ (b) $x \geq -5$ (c) $x \leq 3$ (d) $x < 3$ (e) $x \geq -1$ (f) $x > 7$
(g) $x < -1$ (h) $x < \frac{1}{2}$ (i) $x \leq -1$ (j) $x > 9$ (k) $x \leq -3$ (l) $x > 1\frac{1}{2}$
(m) $x \leq 2\frac{1}{2}$ (n) $x < -\frac{1}{2}$ (o) $x \leq -2\frac{1}{2}$

## Exercise 14.1 (page 157)

1 (a) $\frac{49}{100}$ (b) $\frac{1}{2}$ (c) $\frac{1}{4}$ (d) $\frac{1}{10}$ (e) $\frac{13}{20}$ (f) $\frac{1}{100}$ (g) $\frac{3}{100}$ (h) 1 (i) $2\frac{1}{2}$ (j) $\frac{1}{8}$ (k) $\frac{1}{3}$ (l) $\frac{11}{80}$

2 (a) 0.39 (b) 0.07 (c) 0.3 (d) 1.12 (e) 0.038 (f) 0.625 (g) 0.035 (h) 0.1225

3 (a) 71% (b) 4% (c) 80% (d) 175% (e) 190% (f) 2.7% (g) 3.25% (h) 0.9%

4 (a) 43% (b) 9% (c) 70% (d) 62% (e) 68% (f) 80% (g) 53% (h) 37.5%

5 (a)(i) 0.52 (ii) $\frac{13}{25}$ (b) 48%

6 (a) 55% (b) $\frac{11}{20}$

## Exercise 14.2 (page 159)

1 (a) 18 (b) 12 (c) 27 kg (d) 25 m (e) 9 cm (f) 9 l

2 (a) 24 (b) 115 (c) 24.32 (d) 38 (e) 64.8 (f) 102 (g) 965
(h) 782 kg (i) 124.8 l (j) 2.79 km (k) 19.8 cm (l) 46.5 m

3 (a) 684 (b) 43% (c) 516

4 68

5 792 min

6 (a) £795.80 (b) £332.28 (c) £163.80

7 £1.78

8 (a) £1230 (b) 85%

## Exercise 14.3 (page 160)

1 (a) 78 (b) 436 (c) 873.6 (d) 6384 (e) 78.4 (f) 772.5

2 (a) 49 (b) 387 (c) 228 (d) 2720 (e) 623.2 (f) 36.9

3 754

4 £467.50

5 £2444

6 £11,158.40

7 £658.35

8 £113.75

## Exercise 14.4 (page 161)

1 84%

2 60%, 40%

3 55.3%

4 71%

5 (a) 36% (b) 40% (c) 75% (d) 35.4% (e) 45% (f) 17%
(g) 19.4% (h) 11.7% (i) 11.9%

## Exercise 14.5 (page 163)

1 8%

2 12%

3 4%

4 12.5%

5 7%

6 75%

7 34%

8 28%

## Exercise 14.6 (page 165)

1 (a) 0.69 (b) 1.43 (c) 0.7 (d) 1.07 (e) 1.175 (f) 0.966

2 £169

3 57.8 million

4 £9159.20

5 (a) £554.36 (b) £54.36

6 £750

7 £175

8 £12 500

## Exercise 15.1 (page 167)

1 (a) $A = \frac{1}{2}bh$ (b) 28

2 (a) $s = \frac{d}{t}$ (b) 56

3 (a) $S = 180(n-2)$ (b) 1260

4 (a) $P = 8d$ (b) 29.6

5 (a) $A = l^2$ (b) 73.96

6 (a) $m = \frac{1}{3}(a+b+c)$ (b) 35

7 (a) $A = \frac{1}{2}(a+b)h$ (b) 58

8 (a) $x = \frac{360}{n}$ (b) 40

9 55

10 8

11 −180

12 336

## Exercise 15.2 (page 169)

1 14

2 12

3 2.5

4 7

5 6

6 4

7 126

8 −3

9 5

10 3

11 5 or −5

12 4 or −4

13 10 or −10

14 6 or −6

## Exercise 15.3 (page 171)

In some parts, your answer could take a different form and still be correct.

1 (a) $R = \frac{V}{I}$ (b) $b = \frac{V}{h}$ (c) $h = \frac{E}{mg}$ (d) $y = P - 2x$ (e) $T = \frac{D}{S}$ (f) $h = \frac{3V}{A}$
(g) $R = \frac{100I}{PT}$ (h) $V = \frac{kT}{P}$ (i) $a = p - b - c$ (j) $x = \frac{1}{2}(P - y)$ (k) $u = v + gt$
(l) $a = \frac{v - u}{t}$ (m) $t = \frac{u - v}{g}$ (n) $l = \frac{1}{2}(P - 2b)$ (o) $v = \frac{l + mu}{m}$ (p) $n = \frac{1}{180}(S + 360)$
(q) $a = 2s - b - c$ (r) $a = \frac{2A - bh}{h}$

## Exercise 16.1 (page 176)

1 11.6 cm

2 39.8 m

3 24.2 cm

4 15.4 m

5 40 100 km

6 10.5 m

7 (a) 2.23 m (b) 448

8 31.8 m

## Exercise 16.2 (page 178)

1 2170 cm$^2$

2 616 m$^2$

3 5.64 cm

4 8.59 cm

5 83.9 m$^2$

6 434 mm$^2$

7 80.2 cm$^2$

8 28.3 m$^2$

9 4

10 3180 m$^2$

## Exercise 16.3 (page 180)

1 27°

2 $b = 29°$, $c = 61°$

3 58°

4 68°

5 61°

6 53°

## Exercise 17.1 (page 184)

1 (a) $\frac{29}{50}$ (b) 0.58 (c) 58%

2 (a) $\frac{17}{40}$ (b) $\frac{23}{40}$

3 $\frac{11}{50}$

4 $\frac{1}{5}$

5 $\frac{3}{10}$

## Exercise 17.2 (page 186)

1 (a) $\frac{1}{6}$ (b) $\frac{1}{2}$ (c) $\frac{1}{2}$ (d) $\frac{1}{3}$

2 (a) $\frac{1}{5}$ (b) $\frac{3}{5}$ (c) $\frac{2}{5}$ (d) 0 (e) $\frac{4}{5}$ (f) $\frac{4}{5}$

3 (a) $\frac{3}{10}$ (b) $\frac{7}{10}$ (c) $\frac{7}{10}$ (d) 1

4 (a) $\frac{1}{52}$ (b) $\frac{1}{13}$ (c) $\frac{1}{4}$ (d) $\frac{1}{2}$ (e) $\frac{2}{13}$ (f) $\frac{4}{13}$

5 (a) $\frac{1}{11}$ (b) $\frac{2}{11}$ (c) 0 (d) $\frac{4}{11}$ (e) $\frac{3}{11}$ (f) $\frac{9}{11}$

6 (a) $\frac{1}{500}$ (b) $\frac{1}{100}$

7 $\frac{7}{9}$

8 0.3

## Exercise 17.3 (page 189)

1 (a) $\frac{1}{6}$ (b) $\frac{11}{36}$ (c) $\frac{1}{6}$

2 (a) $\frac{1}{36}$, $\frac{2}{36}$, $\frac{3}{36}$, $\frac{4}{36}$, $\frac{5}{36}$, $\frac{6}{36}$, $\frac{5}{36}$, $\frac{4}{36}$, $\frac{3}{36}$, $\frac{2}{36}$, $\frac{1}{36}$, (b) 7 (c) Sum of probabilities is 1. The sum of the numbers on the dice is certain to be a whole number between 2 and 12 inclusive.

**3** (a) $\frac{1}{12}$ (b) $\frac{1}{6}$ (c) $\frac{1}{4}$ (d) $\frac{1}{4}$

**4** (b)(i) $\frac{2}{9}$ (ii) $\frac{1}{18}$ (c) 1

**5** (a) 6 (b) H1 H2 H3 T1 T2 T3 (d)(i) $\frac{1}{6}$ (ii) $\frac{1}{3}$ (iii) $\frac{1}{2}$

**6** (a) 25 (c)(i) $\frac{4}{25}$ (ii) $\frac{1}{5}$ (d) 6

**7** (a) 18 (b)(i) $\frac{1}{18}$ (ii) $\frac{1}{9}$ (iii) $\frac{1}{9}$

**8** 6

## Exercise 17.4 (page 193)

**1** (b)(i) $\frac{15}{32}$ (ii) $\frac{25}{64}$

**2** (b)(i) $\frac{8}{9}$ (ii) $\frac{5}{9}$

**3** (a) 0.28 (b) 0.18 (c) 0.82

**4** (a) $\frac{1}{4}$ (b) $\frac{1}{169}$ (c) $\frac{1}{32}$ (d) $\frac{1}{26}$ (e) $\frac{7}{16}$

**5** (a) $\frac{2}{27}$ (b) $\frac{4}{27}$ (c) $\frac{19}{54}$

**6** (a) $\frac{4}{81}$ (b) $\frac{29}{81}$ (c) $\frac{40}{81}$

**7** (a) 0.09 (b) 0.04 (c) 0.75

**8** (a) 0.36 (b) 0.01 (c) 0.51

## Exercise 17.5 (page 195)

**1** 250

**2** (a) 50 (b) 100 (c) 150

**3** (a) 125 (b) 75 (c) 175

**4** 520

**5** 12

**6** (a) 50 (b) 100 (c) 150

**7** (a) 60 (b) 120 (c) 220

**8** (a) 25 (b) 75

**9** 10

**10** 50

## Exercise 18.1 (page 199)

**1** (a) 5, 6, 9 (b) 8, 12, 18 (c) 5, 5, 8 (d) 4, 4, 6

**2** (b) 7, 10, 15

**3** 24

**4** regular octahedron (8 faces), regular dodecahedron (12), regular icosahedron (20)

## Exercise 18.2 (page 201)

**1** (b) 294 cm$^2$

**2** 228 cm$^2$

3 (b) 108 cm²

4 (b) 224 cm²

5 351 cm²

6 (b) 391 cm²

## Exercise 18.3 (page 203)

1 729 cm³

2 (a) 160 cm³ (b) 210 m³

3 120 000 cm³

4 1.8 m³

5 3 cm

6 5 cm

7 6721

8 36

9 1 000 000 cm³

10 96

## Exercise 18.4 (page 205)

1 540 cm³

2 480 cm³

3 1390 cm³

4 342 cm³, 1575 cm³

5 8 cm

6 4050 m³

7 864 m³

8 6910 cm³

9 864 cm³

10 3.82 cm

## Exercise 18.5 (page 207)

1 2240 g or 2.24 kg

2 21 758 g or 21.758 kg

3 1210 g or 1.21 kg

4 1347.5 kg

5 53 g

## Exercise 19.1 (page 210)

1 (a) 2:3 (b) 2:1 (c) 4:3 (d) 3:5

2 (a) 4:3 (b) 1:3 (c) 2:5 (d) 5:2 (e) 2:9 (f) 2:5 (g) 10:3 (h) 3:5

3 (a) 2:3 (b) 4:9

4 (a) 1:2 (b) 1:4 (c) 1:8

5 (a) 1:4 (b) 1:1.5 (c) 1:2.25 (d) 1:0.8

6 (a) 6:1 (b) 3.5:1 (c) 2.3:1 (d) 0.75:1

7 16.1:1, 16.4:1, Sinton School.

## Exercise 19.2 (page 211)

1 (a) 3.7 m (b) 1.61 cm

2 (a) 7 m, 6.2 m (b) 16 cm, 13.6 cm

3 (a) 61.92 m (b) 35 cm

4 (a) 3.2 km (b) 20 cm

5 1:25 000

## Exercise 19.3 (page 214)

1 (a) 150 g (b) 60 g (c) 640 g (d) 600 g, 200 g

2 (a) 15 l (b) 20 l

3 36

4 44 cm

5 21 cm, 28 cm

6 £56, £24

7 $120, $600

8 36

9 60°, 50°, 70°

10 100 cm

## Exercise 19.4 (page 215)

1 301 miles

2 $17.55

3 (a) 240 miles (b) 2.5 h

4 (a) 180 m$^2$ (b) 20 l

5 (a) 7.5 (b) 7 h

6 27 g

7 (a) 600 g, 3 kg (b) 450 g, 2.25 kg (c) 200 g, 1 kg

8 (a) 48 kg (b) 24 kg

9 550

10 (a) £12.60 (b) 9524 yen

## Exercise 19.5 (page 217)

1  12 h

2  10 days

3  6 h

4  6 days

5  1 h 10 min

## Exercise 20.1 (page 223)

1  19 cm², 17 cm², 16 cm²

2  Yes. 12 + 18 = 30

3  (a) 6.40 cm (b) 25.6 cm (c) 2.65 cm (d) 13 cm (e) 6 cm
   (f) 3.96 cm

4  7.07 cm

5  12.2 cm

6  12.5 cm

7  5.20 cm

8  24 cm

## Exercise 20.2 (page 225)

1  6.32 m

2  6.71 m

3  51.6 miles

4  10 cm

5  13.9 cm

6  3 cm, 6.40 cm

7  4.24 cm

8  5

## Exercise 20.3 (page 228)

1  (a) 0.839 (b) 2.14 (c) 0.0349 (d) 0.642 (e) 0.0122 (f) 3.69
   (g) 573 (h) 5730

2  (a) 31.0° (b) 71.6° (c) 14.0° (d) 78.7° (e) 4.6° (f) 41.8°
   (g) 83.9° (h) 87.6°

## Exercise 20.4 (page 230)

1  (a) 36.9° (b) 2.04 cm (c) 4.04 cm (d) 24.0° (e) 5.91 cm
   (f) 29.9° (g) 4.47 cm (h) 58.2°

2  9.22 m

3  37.9°

4  11.2 miles

5  301 m

## Exercise 20.5 (page 233)

1  (a) 26.4° (b) 6.71 cm (c) 3.86 cm (d) 31.0° (e) 6.66 cm (f) 27.4° (g) 11.2 cm (h) 6.77 cm

2  62.6 m

3  53.1°

4  6.61 m

5  44.4°

6  47.6 cm$^2$

7  48.2°, 48.2°, 83.6°

8  47.0 cm

## Exercise 21.1 (page 236)

1  (a) $5^2$ (b) $7^3$ (c) $2^4$

2  (a) 9 (b) 125 (c) 16 (d) 1 000 000

3  (a) 3 (b) 4 (c) 4

4  (a) 72 (b) 400 (c) 1620

## Exercise 21.2 (page 237)

1  (a) $2^9$ (b) $5^6$ (c) $3^8$ (d) $10^4$ (e) $5^8$ (f) $2^{12}$ (g) $3^6$ (h) $10^5$

## Exercise 21.3 (page 238)

1  (a) $2^2 \times 3$ (b) $2 \times 3^2$ (c) $2^2 \times 3^2$ (d) $3^2 \times 5$ (e) $2^4 \times 3$
(f) $2^2 \times 3 \times 5$ (g) $2^2 \times 3^3$ (h) $3^2 \times 5^2$ (i) $2^3 \times 5 \times 7$
(j) $2^3 \times 5^2 \times 11$

## Exercise 21.4 (page 239)

1  1 (a) 3 (b) 7 (c) 6 (d) 10

2  (a) $2^2 \times 3^2$ (b) $2^3 \times 3^3$ (c) $2 \times 5^2$ (d) $2^3 \times 5$

3  (a) 24 (b) 60 (c) 40

## Exercise 21.5 (page 240)

1  (a) 15 (b) 12 (c) 24 (d) 12

2  (a) $2^3 \times 3^5$ (b) $2^5 \times 3^2 \times 5 \times 7^3$ (c) $2^6 \times 5^2 \times 11^2$
(d) $2^4 \times 3^3 \times 5^3 \times 7$

3  (a) 200 (b) 120 (c) 540

### Exercise 21.6 (page 242)

1 (a) $9 \times 10^2$ (b) $3 \times 10^3$ (c) $7 \times 10^5$ (d) $4 \times 10^7$ (e) $3.2 \times 10^4$
(f) $1.7 \times 10^5$ (g) $4.52 \times 10^3$ (h) $2.75 \times 10^8$ (i) $5.2 \times 10^7$
(j) $4.6 \times 10^6$

2 (a) 500 (b) 80 000 (c) 90 000 000 (d) 870 000 (e) 9100
(f) 2 300 000 (g) 19 700 (h) 42 100 000 000

3 $3 \times 10^8$

4 $7.5 \times 10^9$

5 93 700 000

### Exercise 21.7 (page 243)

1 (a) $7 \times 10^{-2}$ (b) $4 \times 10^{-4}$ (c) $3 \times 10^{-1}$ (d) $9 \times 10^{-8}$
(e) $8.4 \times 10^{-3}$ (f) $2.9 \times 10^{-5}$ (g) $7.91 \times 10^{-3}$ (h) $7.39 \times 10^{-1}$

2 (a) 0.004 (b) 0.000 08 (c) 0.9 (d) 0.000 000 003 (e) 0.013
(f) 0.000 97 (g) 0.000 004 73 (h) 0.000 000 508

3 $5 \times 10^{-4}$

4 $1.69 \times 10^{-5}$

5 0.000 002 2

### Exercise 21.8 (page 245)

1 (a) $6 \times 10^{11}$ (b) $2.4 \times 10^8$ (c) $3 \times 10^3$ (d) $8 \times 10^3$ (e) $4.3 \times 10^4$ (f) $4.93 \times 10^6$
(g) $3.147 \times 10^5$ (h) $5.92 \times 10^3$

2 (a) $2.88 \times 10^{10}$ (b) $2.55 \times 10^2$ (c) $5.69 \times 10^4$ (d) $9.19 \times 10^7$

3 $2.88 \times 10^{12}$

4 20

5 (a) $2.5 \times 10^{17}$ (b) $1.5 \times 10^6$

### Exercise 22.1 (page 250)

1 (a) 7, 8, 9 (b) 5, 7, 8 (c) 9, 8, 6 (d) 7, 6, 5 (e) 7, 9.5, 10.5
(f) 5, 0, 1

2 3, 2.5, 2.4

3 84, 76, 75

4 1, 1.5, 2.5

5 −1°C, −2°C, −3°C

6 2 h 14 min, 2 h 9 min, 2 h 6 min

7 6 years 1 month

8 (a) 42 mph (b) $2\frac{1}{2}$ h

9 83

10 27

11 1, 1, 5, 6, 7

12  45 kg

13  158 cm

14  18 mph

## Exercise 22.2 (page 251)

1  (a) 7 (b) 25 (c) 7

2  21

3  34, 76

4  9°C

5  3, 3, 8, 9, 12

## Exercise 22.3 (page 253)

1  4, 3, 3.2, 6

2  0, 1, 1.6

3  2, 3, 3, 6

4  15, 15.5, 15.5, 4

## Exercise 22.4 (page 257)

Answers obtained from your cumulative frequency graphs may differ slightly from those given here.

1  (a) 11–15 (b) 11.5 (c) 12, 8, 14, 6

2  (a) $90 \leq w < 100$ (b) 89.1 kg (c) 91 kg, 81 kg, 98 kg, 17 kg
   (d) 85

## Exercise 23.1 (page 262)

1  (a) $y = x + 3$ (b) $y = x + 1$ (c) $y = x - 1$ (d) $x + y = 4$ (e) $x + y = 1$ (f) $x + y = -2$
   (g) $y = 4x$ (h) $y = \frac{1}{2}x$ (i) $y = -2x$

## Exercise 23.2 (page 264)

1  $-8, -5, -2, 1, 4, 7, 10$

2  $-2, -1, 0, 1, 2, 3, 4$

3  $1, -1, -3, -5, -7, -9, -11$

4  (a) (0,3) (b) (0,4) (c) (0,−2) (d) (0,3) (e) (0,2) (f) (0,6)

## Exercise 23.3 (page 268)

1  (a) 7 (b) (0,−4)

2  $y = -5x + 4$

3  $y = 2x + 5, 10.5$

4 (a) 4 (b) $-\frac{1}{2}$ (c) 0

5 (a) $y = 4x + 3$ (b) $y = \frac{1}{2}x + 3$ (c) $y = -5x + 12$

6 (a) $-2, (0,3)$ (b) $\frac{2}{5}, (0,-3)$ (c) $-\frac{3}{4}, (0,6)$

## Exercise 23.4 (page 271)

1 (a) 19, 12, 7, 4, 3, 4, 3, 7, 12, 19 (b) 80, 45, 20, 5, 0, 5, 20, 45, 80 (c) 8, 4.5, 2, 0.5, 0, 0.5, 2, 4.5, 8 (d) $-16, -9, -4, -1, 0, -1, -4, -9, -16$ (e) $-32, -18, -8, -2, 0, -2, -8, -18, -32$ (f) 24, 15, 8, 3, 0, $-1$, 0, 3, 8 (g) 25, 16, 9, 4, 1, 0, 1, 4, 9 (h) 31, 20, 11, 4, $-1, -4, -5, -4, -1$ (i) 15, 4, $-3, -6, -5$, 0, 9, 22, 39

2 (a) $-19, 0, 7, 8, 9, 16, 35$ (b) $-54, -16, -2, 0, 2, 16, 54$ (c) $-13.5, -4, -0.5$, 0, 0.5, 4, 13.5 (d) 27, 8, 1,0, $-1, -8, -27$ (e) 54, 16, 2, 0, $-2, -16, -54$ (f) $-64, -27, -8, -1, 0, 1, 8$ (g) $-36, -12, -2, 0, 0, 4, 18$ (h) $-30, -8, 0$, 0, $-2, 0, 12$ (i) $-38, -16, -6, -2, 2, 12, 34$

3 (a) $-0.7, -1, -2, -4, -10, 10, 4, 2, 1, 0.7$ (b) 0.3, 0.5, 1, 2, 5, $-5, -2, -1$, $-0.5, -0.3$ (c) 0.7, 1, 2, 4, 10, $-10, -4, -2, -1, -0.7$ (d) 4.7, 4.5, 4, 3, 0, 10, 7, 6, 5.5, 5.3

## Exercise 24.1 (page 275)

1 (a) e.g. $x = 2, y = 3$ (b) $x = 3, y = 0$ (c) $x = 3, y = 0$

2 (a), (c)

3 (b)

4 (a) $x = 3, y = 1$ (b) $x = 2, y = 3$ (c) $x = 4, y = 2$ (d) $x = 4, y = -3$ (e) $x = -2$, $y = 4$ (f) $x = -1, y = -2$

5 The lines are parallel so they do not meet.

## Exercise 24.2 (page 278)

1 (a) $x = 5, y = 2$ (b) $x = 2, y = 1$ (c) $x = 3, y = -1$

2 (a) $x = 1, y = 1$ (b) $x = 2, y = -1$ (c) $x = 3, y = 2$ (d) $x = 7, y = -3$ (e) $x = 2\frac{1}{2}$, $y = 2$ (f) $x = \frac{1}{2}, y = -4$

3 49, 25

4 £3

5 39, 83

6 $m = \frac{1}{2}, c = -3$

7 $a = 1, b = -2$

## Exercise 24.3 (page 280)

1 (a) $x^2 + 5x + 6$ (b) $x^2 - x - 20$ (c) $x^2 - 4x - 5$
(d) $x^2 - 9x + 18$ (e) $x^2 + 10x + 25$ (f) $x^2 - 4x + 4$ (g) $x^2 - 49$ (h) $x^2 - 16$
(i) $18 + 3x - x^2$ (j) $5 + 4x - x^2$ (k) $6x^2 + 23x + 20$
(l) $20x^2 - 11x - 3$ (m) $12x^2 - 44x + 7$ (n) $9x^2 + 12x + 4$

(o) $4x^2 - 20x + 25$ (p) $81x^2 - 16$ (q) $9x^2 - 49$ (r) $12 + 28x - 5x^2$ (s) $6 - x - 12x^2$
(t) $25 - 30x + 9x^2$ (u) $a^2x^2 - b^2$

## Exercise 24.4 (page 282)

1  (a) $x + 4$ (b) $x + 2$ (c) $x + 4$ (d) $x - 4$ (e) $x + 8$ (f) $x - 8$ (g) $x - 4$ (h) $x - 2$
   (i) $x + 10$ (j) $x + 2$

2  (a) $(x - 5)(x + 3)$ (b) $(x + 2)(x + 7)$ (c) $(x + 1)(x - 25)$
   (d) $(x + 2)(x - 12)$ (e) $(x + 7)(x - 7)$ (f) $x(x + 8)$ (g) $(x - 4)(x + 14)$
   (h) $(x - 1)(x + 1)$ (i) $(x - 10)(x + 2)$ (j) $x(x - 9)$

## Exercise 24.5 (page 283)

1  (a) $2.65, -2.65$ (b) $0, 2$ (c) $1, 3$ (d) $1.3, -2.3$ (e) $2.8, -1.3$
   (f) $3.6, -0.6$

2  (a) $x = 2$ graph touches $x$-axis but does not cross it (b) no solution as graph
   does not cross the $x$-axis

## Exercise 24.6 (page 285)

1  (a) $1, 6$ (b) $-4, 3$ (c) $-7, 7$ (d) $-5, -4$ (e) $0, 6$ (f) $0, -8$

2  (a) $1, 2$ (b) $-2, 1$ (c) $-4, -3$ (d) $-5, 5$ (e) $5$ repeated
   (f) $0, -10$ (g) $-4, 5$ (h) $-8, 8$ (i) $0, 7$ (j) $-2, 5$

3  $7\,\text{cm}$

4  $9$ or $-3$

5  $8$ or $-9$

6  $7, 9$ (or $-7, -9$)

7  $7, 8$

## Exercise 24.7 (page 288)

1  (a) $0, 2, 5$ (b) $-2, 1, 3$ (c) $-2.2, -0.8, 2.9$ (d) $-2, 1, 2$
   (e) $0, 2$ repeated (f) $2.4$

2  $5.8$

3  $4.3$

4  $2.9$

5  $1.7$

6  $2.3$

**taking it further**

Teach Yourself Mathematics is a general course that covers and integrates different branches of mathematics. To go on from here, you could either pursue a more advanced general course in mathematics, or decide that you would like to know more about the individual subjects, algebra, trigonometry, calculus and statistics which make up Teach Yourself Mathematics. Whichever line you decide to take, you should think carefully about buying a graphical display calculator if you do not have one already. These calculators are not cheap, but they are important tools for learning trigonometry, calculus and statistics. Recommended graphical display calculators at the time of writing (2008) are Texas Instruments TI84 and Casio 9850+.

## Websites and organizations

A + B Books, who write and publish mathematics books for use with a graphics calculator.
**www.AplusB.co.uk**

Association of Teachers of Mathematics (UK)
**www.atm.org.uk**

Autograph: software for learning advanced mathematics
**www.autograph-math.com**

Bournemouth University applets (computer animations) demonstrating mathematical principles
**mathinsite.bmth.ac.uk**

Coventry University Mathematics Support Centre for Mathematics Education
**www.mis.cov.ac.uk/maths_centre**

Free-Standing Mathematics Qualifications
**www.fsmq.org**

International Baccalaureate: an international qualification
**www.ibo.org**

iCT Training Centre, with many useful resources and links to other sites worldwide
**www.tsm-resources.com**

Mathematical Association (UK)
**www.m-a.org.uk**

Mathpuzzle contains a range of puzzles and other maths resources
**www.Mathpuzzle.com**

MathsNet contains a wide range of puzzles, books and software
**http://www.mathsnet.net**

National Council of Teachers of Mathematics (USA)
**www.nctm.org**

Open University Mathematics courses
**www.mathematics.open.ac.uk**

Oundle School site, with many useful resources and links *to* other sites worldwide
**http://www.argonet.co.uk/oundlesch/mlink.html**

National Council of Teachers of Mathematics (USA)
**www.nctm.org**

UK National Statistics on line
**http ://www. statistics.gov.uk**

University of Plymouth Mathematics support Materials
**www.tech.plym.ac.uk/maths/resources/PDFLaTeX/mathaid. html**

# Reading list

Abbott, P. (2003) *Teach Yourself Algebra*, Hodder Education, London.

Abbott, P. (2003) *Teach Yourself Trigonometry*, Hodder Education, London.

Abbott, P. (2003) *Teach Yourself Calculus*, Hodder Education, London.

Barrow, John D. (1993) *Pi in the Sky: Counting, Thinking and Being*, Penguin, London. An exploration of where maths comes from and how it is performed.

Eastaway, Rob & Wyndham, Jeremy (1998) *Why Do Buses Come in Threes?*, Robson Books, London. Practical uses for various mathematical topics, including probability, Venn Diagrams and prime numbers.

Eastaway, Rob & Wyndham, Jeremy (2002) *How long is a piece of string?*, Robson Books, London. Examples of mathematics in everyday life.

Eastaway, Rob & Wells, David (2005) *Mindbenders and Brainteasers*, Robson Books, London. A collection of 100 puzzles and conundrums, old and new.

Flannery, Sarah (2000) *In Code: A Mathematical Journey,* Profile Books, London. A collection of problems with solutions and explanations, based on the author's experiences of growing up in a mathematical home.

Graham, Alan (2003) *Teach Yourself Statistics*, Hodder Education, London. A straightforward and accessible account of the big ideas of statistics with a minimum of hard mathematics.

Haighton, June et al., (2004) *Maths the Basic Skills*, Nelson Thornes, Cheltenham. A traditional textbook on basic mathematics based on the Adult Numeracy Core Curriculum.

Huntley, H.E. (1970) *The Divine Proportion, Study in Mathematical Beauty,* Dover, New York. Applications in art and nature of the 'Golden Ratio'.

Ifrah, Georges (1998) *The Universal History of Numbers,* The Harvill Press, London. A detailed book (translated from French) about the history of numbers and counting from prehistory to the age of the computer.

Neill, H. and Quadling, D. (2007) *Mathematics for the IB Diploma*, Standard Level, Cambridge University Press.

Paulos, John Allen (1990) *Innumeracy – Mathematical Illiteracy and its Consequences*, Penguin, London. Real-world examples of innumeracy, including stock scams, risk perception and election statistics.

Pólya, G. (1990) *How to Solve It*, Penguin, London. A classic text on mathematical problem solving that is well-known around the world.

Potter, Lawrence (2006) *Mathematics Minus Fear*, Marion Boyars Publishers Ltd, London. A romp through school mathematics that takes in puzzles and gambling.

Seiter, Charles (1995) *Everyday Math for Dummies*, For Dummies, New York. A friendly guide to basic maths, written in an entertaining style.

Singh, Simon (2000) *The Code Book*, Fourth Estate, London. A history of codes and ciphers and their modern applications in electronic security.

Singh, Simon (1998) *Fermat's Last Theorem*, Fourth Estate, London. An account of Andrew Wiles' proof of Fermat's Last Theorem, but also outlining some problems that have interested mathematicians over many centuries.

Stewart, Ian (1996) *From Here to Infinity*, Oxford University Press, Oxford. An introduction to how mathematical ideas are developing today.

Stewart, Ian (1997) *Does God Play Dice?*, Penguin, London. An introduction to the theory and practice of chaos and fractals.

Stewart, Ian (2006) *Letters to a Young Mathematician,* Basic Books, New York. What the author wishes he knew about mathematics when he was a student.

# index